The Best American Science and Nature Writing 2006

GUEST EDITORS OF
THE BEST AMERICAN SCIENCE
AND NATURE WRITING

2000 DAVID QUAMMEN
2001 EDWARD O. WILSON
2002 NATALIE ANGIER
2003 RICHARD DAWKINS
2004 STEVEN PINKER
2005 JONATHAN WEINER
2006 BRIAN GREENE

The Best American Science and Nature Writing 2006

Edited and with an Introduction by Brian Greene

Tim Folger, Series Editor

HOUGHTON MIFFLIN COMPANY
BOSTON · NEW YORK 2006

Visit our Web site: www.houghtonmifflinbooks.com.

ISSN: 1530-1508
ISBN-13: 978-0-618-72221-1 ISBN-10: 0-618-72221-1
ISBN-13: 978-0-618-72222-8 (pbk.) ISBN-13: 0-618-72222-X (pbk.)

Printed in the United States of America

MP 10 9 8 7 6 5 4 3 2

"The Blogs of War" by John Hockenberry. First published in *Wired,* August 2005. Copyright © 2005 by John Hockenberry. Reprinted by permission of the author and the Watkins/Loomis Agency.

"The Forgotten Era of Brain Chips" by John Horgan. First published in *Scientific American,* October 2005. Copyright © 2005 by John Horgan. Reprinted by permission of the author.

"The Mysteries of Mass" by Gordon Kane. First published in *Scientific American,* July 2005. Copyright © 2005 by Scientific American, Inc. Reprinted by permission of Scientific American, Inc.

"Future Shocks" by Kevin Krajick. First published in *Smithsonian,* March 2005. Copyright © 2005 by Kevin Krajick. Reprinted by permission of the author.

"The Mummy Doctor" by Kevin Krajick. First published in *The New Yorker,* May 16, 2005. Copyright © 2005 by Kevin Krajick. Reprinted by permission of the author.

"X-Ray Vision" by Robert Kunzig. First published in *Discover,* February 2005. Copyright © 2005 by Robert Kunzig. Reprinted by permission of the author.

"The Illusion of Gravity" by Juan Maldacena. First published in *Scientific American,* November 2005. Copyright © 2005 by Scientific American, Inc. Reprinted by permission of Scientific American, Inc.

"The Coming Death Shortage" by Charles C. Mann. First published in *The Atlantic Monthly,* May 2005. Copyright © 2005 by Charles C. Mann. Reprinted by permission of the author.

"The Dover Monkey Trial" by Chris Mooney. First published in *Seed,* October/November 2005. Copyright © 2005 by Chris Mooney. Reprinted by permission of the author.

"Remembrance of Things Future" by Dennis Overbye. First published in the *New York Times,* December 12, 2005. Copyright © 2005 by the New York Times Co. Reprinted by permission of the New York Times Co.

"Out of Time" by Paul Raffaele. First published in *Smithsonian,* April 2005. Copyright © 2005 by Paul Raffaele. Reprinted by permission of the author.

"Torrential Reign" by Daniel Roth. First published in *Fortune,* October 31, 2005. Copyright © 2005 by Time, Inc. Reprinted by permission of Time, Inc.

"Are Antibiotics Killing Us?" by Jessica Snyder Sachs. First published in *Discover,* November 2005. Copyright © 2005 by Jessica Snyder Sachs. Reprinted by permission of the author.

"Remembering Francis Crick" by Oliver Sacks. First published in the *New York Review of Books,* March 24, 2005. Copyright © 2005 by Oliver Sacks. Reprinted by permission of the Wylie Agency, Inc.

"Buried Suns" by David Samuels. First published in *Harper's Magazine,* June 2005. Copyright © 2005 by David Samuels. Reprinted by permission of the author.

"Lights, Camera, Armageddon" by Joshua Schollmeyer. First published in *Bulletin of the Atomic Scientists,* May/June 2005. Copyright © 2005 by *Bulletin of the Atomic Scientists.* Reprinted by permission of *Bulletin of the Atomic Scientists.*

"Taming Lupus" by Moncef Zouali. First published in *Scientific American,* March 2005. Copyright © 2005 by Scientific American, Inc. Reprinted by permission of Scientific American, Inc.

Contents

Foreword

In February 1676, when Isaac Newton was at the height of his fame, he wrote the following to Robert Hooke, a notoriously testy colleague and envious rival: "If I have seen farther than others, it is because I was standing on the shoulders of giants." Call me small-minded, but I doubt the sincerity of Newton's sentiment; he was by all accounts a petulant and vindictive genius. But leaving aside the question of motivation, I think what Newton really did was kick the giants in the shins and then pull the rug out from under the whole tottering tower of behemoths.

Because that's what scientists do — they try their utmost to knock down the best efforts of their fellows. And they love it when they succeed. If it's someone's life work that is savaged, or the foundation of an entire discipline, so much the better. The late philosopher of science Karl Popper called the process falsification, meaning that science advances by proving existing theories wrong. I call it chum in the water, because sooner or later even the most successful theory draws a frenzied and merciless attack.

All of which makes me think that the root meaning of the word "science" — it comes from the Latin *scire,* "to know" — kind of misses the point. Closer to the true contentious, ornery spirit of science would be an etymology for "I know you're wrong." Seen from a distance, scientific research looks orderly, austere, perhaps even a bit sterile. Seen from up close, it's an intensely human endeavor, with nobility and self-sacrifice commingling with self-doubt, ambition, jealousy, swollen egos, and sometimes outright fraud. Last December, for example, Hwang Woo Suk, a South Korean scientist who had made headlines around the world by claiming to have cloned human stem cells, was forced to resign from his university in disgrace when it was discovered that he had fabricated his research results.

Yet somehow, something remarkable emerges from all the tumult. Even though the intellectual brawls never stop, charlatans are invariably exposed, and the ceaseless, collective, rigorous drive to find fault yields an understanding of reality impossible to achieve by any other means. All the arguments, hunches, conferences, particle accelerators, papers, data, and flashes of insight come together to give us a whole far greater than the sum of its flawed human parts: a vision of a universe whose size, richness, and strangeness no one of us alone could have imagined. Given the means, the fantastic ends seem almost undeserved. They're certainly unlikely. The entire manic enterprise reminds me of the construction of an anthill, without the cooperation.

The human face of science is on full display in this collection, and so is the shin-kicking assault on established theories. For an example of the former, I direct you to Daniel Roth's "Torrential Reign," a profile of Bram Cohen, the thirty-one-year-old creator of the BitTorrent file-sharing software that is roiling the Internet. A few years ago Cohen, who has a mild form of autism, was unemployed and lived on credit cards; today . . . But read it yourself — I've already given away too much.

As for the shin-kicking, consider "The Illusion of Gravity," in which Juan Maldacena pulls the rug out from under Newton — and all the rest of us. Until now you've probably felt fairly safe in assuming that you live in a three-dimensional world (okay, four, for the general relativists among you). But Maldacena, one of the world's leading theoretical physicists, argues that the universe we see is essentially an illusory hologram. Maldacena, by the way, must be the only physicist in the world to have inspired his colleagues to dance in the aisles at a major conference. You won't read about this in his article, but a few years ago two hundred or so physicists line-danced the macarena, chanting as a refrain, "Ehhh! Maldacena!"

The remaining twenty-three stories are no less remarkable, from Dennis Overbye's report on the world's first conference for time travelers to Paul Raffaele's gripping account of the Korubo people of the Amazon, who for all their fierceness may not survive their encounter with modernity. In your hands now is a sampler, a few elegant takes from a single year in the phenomenon called science.

By the time you read these words, 2006 will be nearly over. I hope readers, as well as writers and editors, will nominate their favorite articles for next year's anthology at http://www.timfolger.net/forums.

I also encourage readers to use my Web site's forums to give me feedback about the collection and to discuss current science-related issues. As an incentive to enlist readers to scour the nation in search of good science and nature writing, I have a little proposition: if you submit an article to my Web site that I haven't found and the article makes it into the anthology, I'll mail you a free copy of next year's *Best American Science and Nature Writing*. I try to read widely, but I live in a small town in a remote part of New Mexico, and I know I'm missing some good material, so send me your favorites. All submissions must follow a few ground rules: They must have been published in the United States during the 2006 calendar year and must be nonfiction articles about science or the environment. No poetry or book excerpts, but essays are welcome. The articles can be from magazines, newspapers, or the Web. I'll forward about one hundred of the best articles I've read to the guest editor of the series, who will choose the final twenty-five or so.

I'll be moving later this year, and as I write, I'm not sure what my mailing address will be when this book is published. So the best way to submit articles is to post links to them on my Web site or to leave a message for me on one of the site's forums, where I will also list my new address. If you send a hard copy, please include the entire publication, tear sheets, or a high-quality photocopy of the original that clearly shows the author's name, the publication date, and the name of the publication. All submissions must reach me by early January 2007. The best way for publications to guarantee that all their articles are considered for inclusion in the anthology is to place me on their subscription list.

I feel privileged to have had the chance to work with Brian Greene, a leading physicist and a superb science writer. I trust he has recovered from the shock of receiving a few pounds of articles shortly before he set sail for Antarctica. Once again I'm indebted to Amanda Cook and Will Vincent, without whom this anthology wouldn't exist and who now realize, I hope, that when it comes to libraries, the Hub of the universe has nothing on Gallup, New Mexico. And endless thanks to my beauteous wife, Anne Nolan, who some years ago pulled the rug out from under me. I'm still reeling.

TIM FOLGER

Introduction

There was a time, long ago, when I wasn't much into words. Books rarely lit a fire in me as a kid. When we were assigned the usual canon of great works in school — from *Macbeth* to *Moby Dick* — I'd diligently start in but, truth be told, I often wouldn't finish. From newspapers to novels, magazines to plays, comics to biographies — the written word was a burden.

Mathematical equations were a completely different story. I loved their precision, their unwavering certainty, and the way they just plain worked. Whether you organized a calculation one way or chose to attack it from a different angle entirely, as long as you executed each step correctly you'd get the same answer. All roads inevitably led to an identical result, with no room for opinion or interpretation. Mathematics allowed me an abundance of creativity in seeking a problem's solution but constrained that creativity to serve a single ultimate purpose: getting the right answer.

As I got a little older, my tastes became more nuanced. It's not that my devotion to the exactitude of mathematics diminished. Rather, I began to better appreciate the gray areas of life and literature. The gray areas of ambiguity, the gray areas of human paradox. The gray areas where, search as you might, you will never find a solution. I spent increasing amounts of time wrapped up with Balzac, Gorky, Faulkner, Orwell, Sartre, and Camus. A messy and wonderful world opened up for me, and the burden of words lifted.

Even so, the two realms — the humanities and the sciences — seemed thoroughly separate, a view that remains widely held. In fact, the divide between the "two cultures" runs even deeper now. As Nicholas Kristof emphasized in a recent *New York Times* op-ed, the "hubris of the humanities" is extensive. There is an implicit agreement in "educated circles" that it's "barbaric to be unfamiliar with Plato or Monet or Dickens, but quite natural to be oblivious of quarks and chi-squares." As a professional scientist, I've often encountered this attitude among nonscientists. It's rarely derogatory, and it's frequently accompanied by embarrassment — sometimes feigned — that the otherwise intelligent and informed individual has no understanding of science or mathematics. Generally, the encounters end with a well-meaning chuckle (one in which decorum obliges me to partake) that says in short "it's really okay not to know any math or science."

But in the twenty-first century, such willful ignorance of science is not okay. We are living through a radical cultural shift, one in which science and technology play an increasingly pervasive role in everyday life. From stem cells to global warming, from cyberspace to nanotechnology, from computer-enhanced perception to extension of the human lifespan, full participation in the global conversation requires a familiarity with the major advances in science and technology as well as an understanding of the scientific way of thought. The most far-reaching choices we will make in the years ahead, whether through action or apathy, are ones that have a critical scientific component. How should we deal with pandemic threats? What limits, if any, should we legislate on genetic manipulation? Should we curtail research on human cloning? Should we support manned space exploration, or is money better spent on robotic missions? Is evolution "just" a theory? Informed decision making requires a populace engaged with science, not one that is looking in from the periphery and not one that takes pride in its lack of knowledge.

Even on issues that seemingly bear no direct relation to science, a scientific mindset can have a radical impact. Here's a concrete example. Think back to the 2000 presidential election and recall the dominant question asked in the tumultuous weeks following the casting of ballots: who received more votes in Florida? Count followed recount, with the vote differentials ranging from a handful to a few dozen to a few hundred. All attention was sharply focused on who came out on top after each of these recounts. But to someone familiar with scientific analysis, these recounts raised a different issue. Every measurement has a built-in limit on its accuracy. With a good-quality bathroom scale, you might measure your weight to an accuracy of half a pound — which for most purposes is all you need. But if you want to determine which of two average-size apples is heavier, your bathroom scale would prove inadequate, as its accuracy is too coarse to provide a meaningful answer. Scientists are acutely aware of such limitations and always accompany their measurements with a caveat declaring the numerical limits on the accuracy of their methods and apparatus.

Vote-counting procedures also have built-in limitations on accuracy. In Florida these inaccuracies arose from hanging chads, dimpled chads, pregnant chads, butterfly ballots, computer malfunctions, errors with voter registration, and other factors. Usually such sources of inaccuracy can be ignored because they're too small to have any effect on the election's outcome. But when the vote differentials are not in the hundreds of thousands or in the tens of thousands or even in the thousands, the built-in inaccuracies compromise the entire process. Like trying to use your bathroom scale to find the heavier dust mote, the vote-counting apparatus is just too coarse to determine a winner. Thus the appropriate question should not have been "who received more votes?" because the sources of error were as large as or larger than the vote differentials, rendering it impossible to provide an answer with any meaning. Instead, the mostly unasked yet primary question was "what should we do when two candidates, to within the accuracy of our measuring apparatus, are tied?"

I offer no answer to this question because clearly that's not my point. Instead, the example highlights how a scientifically sophisticated public would have shifted the debate to the truly relevant question and not bothered with, nor accepted the outcome of, a procedure that was in effect meaningless.

A scientifically literate public is, plainly, increasingly vital.

How then do we make headway toward this end? Well, let's turn the issue around and ask why Kristof's hubris of the humanities isn't met by an equally extensive "smugness of the sciences." Why isn't it considered barbaric to know little or nothing about probability and statistics, genetics and biochemistry, relativity and quantum mechanics? One can come up with many explanations — science requires a specialized language that relatively few people have the inclination to study, science deals with esoteric questions that only experts can truly appreciate, science operates in an abstract, foreign realm that doesn't cater to the casual visitor — but I think all such propositions add up to this: to many people, science — unlike Shakespeare, Beethoven, or Monet — seems dry, cold, and removed from the human spirit. Shakespeare, Beethoven, and Monet tell us about ourselves. They augment our understanding of the range and texture of human experience and thus enrich our limited time here on Earth. Science, many feel, doesn't do this and hence is dispensable or, more precisely, can be safely left to the scientists. One culture is thus deemed profoundly relevant while the other profound but extremely distant.

This line of thought isn't unreasonable. I can see where it comes from and how aspects of our educational system foster it. But the conclusion is wrong and, in the long run, dangerous.

Science needs to be recognized for what it is: the ultimate in adventure stories. Against staggering odds, a species that has been walking upright for only a few dozen millennia is trying to unravel mysteries that have been billions of years in the making. How did the universe begin? How was life initiated? How did consciousness emerge? These are big questions, and the stakes are high. Fully answering these questions — something yet to be achieved, though researchers worldwide are closing in at an exhilarating pace — will provide the very frame for life as we know it and prepare us for the next step, in which we go from being inquirers about nature to becoming its manipulators. In fact, today's researchers routinely tinker with the molecular structure of life, so it's clear that we have already taken the first steps on this latter journey.

My secondary-school education revealed nothing of this drama. The human element of searching into the unknown — the human struggle to piece together a coherent story of life and the universe — played essentially no role in *any* of my science classes. Instead, we focused exclusively on the results. Newton's laws were taught without Newton. Mendeleev's periodic table was taught without Mendeleev. Darwin's evolutionary theory was taught without Darwin. Frankly, I didn't mind this at all. I relished the crispness, the unencumbered clarity, the purity of science. But my reaction has little relevance — I was hooked on math and science by the time I was six. The larger truth is that if my education was at all typical, it is no wonder that science feels removed from the human experience. More often than not, science education removes the humans.

The very nature of mathematics and science compounds the problem, for their level of detail and precision goes well beyond that of other subjects. Grasping these disciplines thus requires that students attain a level of analytical thought that is not only unfamiliar but also almost completely lacking in opportunities for emotional connection — thus perpetuating the myth that the sciences, especially the hard sciences, are cold and aloof.

Some science textbooks do give a nod to historical context, but sidebars and the occasional boxed human-interest interlude have little lasting impact. I've recently become aware of some specialized schools in which science is taught largely through experimentation and hands-on discovery. Students in those schools may well emerge with a very different and more accurate sense of what science is all about. In the vast majority of schools, though, teachers are expected to cover so much material that they have little choice but to focus directly and exclusively on the science proper.

I believe the beginnings of a remedy for this problem are to be found in the realm of popular science writing. If it is well done, nontechnical science writing can connect the sciences and the humanities. And this connection is essential. Like master chefs, the best science writers pare away all but the most succulent material, trimming details essential to the researcher that would be only a distraction to the reader. And by carefully crafting a narrative and using expository devices that showcase the drama of scientific exploration and discovery, popular works can maintain a high level of scientific integrity while making difficult and technical subjects not only accessible but moving and compelling. Good science writing can humanize the abstractions of scientific research by establishing visceral, meaningful connections to questions and issues we care about and by humanizing the scientific process itself. In Einstein's words, scientific research consists of "years of anxious searching in the dark for a truth one feels but cannot find, until a final emergence into the light." A reader who is led to envisage this search, I believe, will start to bridge the gulf between the science and the humanities. The best science writing can have that effect.

I first became aware of the illuminating capacity of science writing during my freshman year in college when I took Robert Nozick's course Introduction to the Problems of Philosophy. Nozick's lectures were inspirational — clear, concise, and provocative. The readings, by contrast, were at times dense to the point of being impenetrable. The philosophers we studied were, by reputation, among the most influential thinkers our species has produced. But as I read some of their works I wondered, does anyone really know what they were trying to say? To be sure, not all the assignments were opaque, but frequently I had to read passages over and over again — guessing here, straining there. At times the level of frustration was severe. In my mathematics and physics courses, I was no stranger to wrestling with difficult, subtle concepts. But spending hours poring over philosophy texts and being uncertain whether I'd made any progress at all — *that* was unfamiliar and disheartening. When I recounted my difficulties to the teaching assistant, himself a graduate student in philosophy, he replied, "Ah . . . struggling over the meaning of another's text, you're doing well — that's the very nature of philosophy."

But then I came upon an essay by Daniel Dennett dealing with the famous mind/body problem, which we were studying in class. The essay was a breath of fresh air. Not only was it thoroughly readable, it was entertaining. And throughout, Dennett stayed true to his material. He laid out the problem of personal identity — does our sense of self come from and reside solely in our brains, or does it somehow transcend that three-pound gray mass? — with clarity and verve through a story he created for the purpose (a tale involving, among other things, his brain being delicately removed from his head yet still in control of his body through a radio-controlled hookup). He went on to describe possible philosophical solutions, which he illustrated through narrative vignettes that made their meaning — as well as their strengths and weaknesses — crystal clear. Whereas some of the other philosophers seemed engaged in a battle with language — I could almost imagine their drafts marked up with change upon change as they wrenched the sentences into hard-won paragraphs — Dennett wielded language as an essential and effective tool.

Although this was not my first encounter with such a "popularization" — I was well aware of Carl Sagan's *Cosmos,* largely from its public television adaptation, and I read *Scientific American* now and then — it was the first that impressed me with the power of the genre. On reflection, I find that this makes perfect sense. Because I was studying math and physics with the intention of pursuing those fields professionally, popularizations in those areas, while enjoyable, didn't serve any particular purpose. Philosophy, though, was a field I was dabbling in just to get a sense of the subject; I had no inclination to study it in depth. Dennett's essay thus hit the mark perfectly: he brought a subtle subject down to earth and did so in an essay that was not only easy but pleasurable to read. That, to me, is the hallmark of excellent science writing.

Over the past few years, as I've gone around the country giving both general and technical lectures at high schools and colleges, I've frequently been asked to offer suggestions for raising the level of interest in science and mathematics. I generally rephrase the question, because my experiences suggest a slightly different perspective. Across the country a great many students *are* interested in science. The task is not so much in attracting them to science, but rather in not turning them off from it. To this end, I've been advocating that schools introduce a new course, one in which students spend the entire term reading and discussing a wide selection of compelling popular-science books and articles.

What would this accomplish? Well, those students who lose interest in science and mathematics generally do so because the material gets difficult and requires more effort to understand than they can — or think they can — expend. This is not a catastrophe. By no means is everyone cut out to be a scientist or mathematician, or even to delve into the complexities of these subjects. But the rigorous, detailed, abstract elements, while critical for those who seek facility in science, perhaps in preparation for a career, are not required for basic science literacy. Having students read gripping, nontechnical materials that cut a wide swath through the sciences, from their established underpinnings up to cutting-edge research, can cultivate excitement while greatly deepening their working knowledge of science's achievements. No doubt if more students saw science as a great venture into the unknown, as an unfolding story of exploration, then more of them would be drawn to participate in their generation's chapter of discovery. At the very least, these students would be more inclined to continue following science in popular sources throughout their lives.

The challenge for popular science writers, of course, is to humanize the subject matter while staying true to the science. The danger is that in highlighting the drama and easing off on the details, the science may be reduced to a mere caricature of the underlying ideas and results. Avoiding this pitfall is the art of the science writer's trade. And, as with all arts, there is no "right" way, no standard approach, and no agreed-upon measure of success. I've read popularizations that I thought were truly wonderful, only to find other readers who thought there was too much science and others still who thought there was too little. I've also read some popularizations that, for my taste, focused too heavily on the human-interest story behind a discovery, but encountered other readers who found the narrative gripping. That is an inevitable — and welcome — result of mixing the messy grays of the humanities with the purity of scientific results.

The selection of articles in this collection reflects the wide range of styles and degrees of scientific emphasis in science writing today. Some convey cutting-edge research in fields ranging from particle physics to neurobiology. Other articles focus on medical issues, from unintended consequences of antibiotics to the critical need in medical education to conduct more autopsies. A number of others are driven by the effects of science and technology on people, including a moving exploration of brain implants, the impact of frontline bloggers on the perception of war, and the wonders of psychopharmacology. A couple of articles cover human development, from aspects of language acquisition to first contact with remote cultures untouched by the modern world, while a couple of others take on the evolution/intelligent-design pseudo-debate.

Collectively, the articles demonstrate the long reach of science's tentacles, a reach that goes well beyond the confines of the ivory tower. They also demonstrate the high level of science writing found in a range of newspapers, magazines, and periodicals, aimed at a wide spectrum of viewers. If we harness this talent by integrating these kinds of articles into the standard secondary education curriculum, the long-range impact on science awareness and science literacy would be of great consequence.

BRIAN GREENE

The Best American Science and Nature Writing 2006

NATALIE ANGIER

Almost Before We Spoke, We Swore

FROM *The New York Times*

INCENSED BY WHAT IT SEES as a virtual pandemic of verbal vulgarity issuing from the diverse likes of Howard Stern, Bono of U2, and Robert Novak, the United States Senate is poised to consider a bill that would sharply increase the penalty for obscenity on the air.

By raising the fines that would be levied against offending broadcasters some fifteenfold, to a fee of about $500,000 per crudity broadcast, and by threatening to revoke the licenses of repeat polluters, the Senate seeks to return to the public square the gentler tenor of yesteryear, when seldom were heard any scurrilous words, and famous guys were not foul mouthed all day.

Yet researchers who study the evolution of language and the psychology of swearing say that they have no idea what mystic model of linguistic gentility the critics might have in mind. Cursing, they say, is a human universal. Every language, dialect, or patois ever studied, living or dead, spoken by millions or by a small tribe, turns out to have its share of forbidden speech, some variant on comedian George Carlin's famous list of the seven dirty words that are not supposed to be uttered on radio or television.

Young children will memorize the illicit inventory long before they can grasp its sense, said John McWhorter, a scholar of linguistics at the Manhattan Institute and the author of *The Power of Babel*, and literary giants have always constructed their art on its spine. "The Jacobean dramatist Ben Jonson peppered his plays with fackings and 'peremptorie Asses,' and Shakespeare could hardly quill a stanza without inserting profanities of the day like 'zounds' or 'sblood' — offensive contractions of 'God's wounds' and 'God's blood' — or some wondrous sexual pun." The title *Much Ado About Nothing,* McWhorter said, is a word play on "Much Ado About an O Thing," the O thing being a reference to female genitalia.

Even the quintessential Good Book abounds in naughty passages, like the men in II Kings 18:27 who, as the comparatively tame King James translation puts it, "eat their own dung, and drink their own piss."

In fact, said Guy Deutscher, a linguist at the University of Leiden in the Netherlands and the author of *The Unfolding of Language: An Evolutionary Tour of Mankind's Greatest Invention,* the earliest writings, which date from five thousand years ago, include their share of off-color descriptions of the human form and its ever-colorful functions. And the written record is merely a reflection of an oral tradition that Deutscher and many other psychologists and evolutionary linguists suspect dates from the rise of the human larynx, if not before.

Some researchers are so impressed by the depth and power of strong language that they are using it as a peephole into the architecture of the brain, as a means of probing the tangled, cryptic bonds between the newer, "higher" regions of the brain in charge of intellect, reason, and planning and the older, more "bestial" neural neighborhoods that give birth to our emotions.

Researchers point out that cursing is often an amalgam of raw, spontaneous feeling and targeted, gimlet-eyed cunning. When one person curses at another, they say, the curser rarely spews obscenities and insults at random, but rather will assess the object of his wrath and adjust the content of the "uncontrollable" outburst accordingly.

Because cursing calls on the thinking and feeling pathways of the brain in roughly equal measure and with handily assessable fervor, scientists say that by studying the neural circuitry behind it they are gaining new insights into how the different domains of the brain communicate — and all for the sake of a well-venomed retort.

Other investigators have examined the physiology of cursing, how our senses and reflexes react to the sound or sight of an obscene word. They have determined that hearing a curse elicits a literal rise out of people. When electrodermal wires are placed on people's arms and fingertips to study their skin conductance patterns and the subjects then hear a few obscenities spoken clearly and firmly, participants show signs of instant arousal. Their skin conductance patterns spike, the hairs on their arms rise, their pulse quickens, and their breathing becomes shallow.

Interestingly, said Kate Burridge, a professor of linguistics at Monash University in Melbourne, Australia, a similar reaction occurs among university students and others who pride themselves on being educated when they listen to bad grammar or slang expressions that they regard as irritating, illiterate, or déclassé. "People can feel very passionate about language," she said, "as though it were a cherished artifact that must be protected at all cost against the depravities of barbarians and lexical aliens." Burridge and a colleague at Monash, Keith Allan, are the authors of *Forbidden Words: Taboo and the Censoring of Language,* which will be published early next year by the Cambridge University Press.

Researchers have also found that obscenities can get under one's goose-bumped skin and then refuse to budge. In one study, scientists started with the familiar Stroop test, in which subjects are flashed a series of words written in different colors and are asked to react by calling out the colors of the words rather than the words themselves. If the subjects see the word "chair" written in yellow letters, they are supposed to say "yellow."

The researchers then inserted a number of obscenities and vulgarities in the standard lineup. Charting participants' immediate and delayed responses, the researchers found that, first of all, people needed significantly more time to trill out the colors of the curse words than they did for neutral terms like "chair."

The experience of seeing titillating text obviously distracted the participants from the color-coding task at hand. Yet those risqué interpolations left their mark. In subsequent memory quizzes, not only were participants much better at recalling the naughty words than they were the neutrals, but that superior recall also applied to the tints of the tainted words, as well as to their sense.

Yes, it is tough to toil in the shadow of trash. When researchers in another study asked participants to quickly scan lists of words that included obscenities and then to recall as many of the words as possible, the subjects were, once again, best at rehashing the curses — and worst at summoning up whatever unobjectionable entries happened to precede or follow the bad bits.

Yet as much as bad language can deliver a jolt, it can help wash away stress and anger. In some settings, the free flow of foul language may signal not hostility or social pathology, but harmony and tranquility. "Studies show that if you're with a group of close friends, the more relaxed you are, the more you swear," Burridge said. "It's a way of saying: 'I'm so comfortable here I can let off steam. I can say whatever I like.' "

Evidence also suggests that cursing can be an effective means of venting aggression and thereby forestalling physical violence. With the help of a small army of students and volunteers, Timothy B. Jay, a professor of psychology at Massachusetts College of Liberal Arts in North Adams and the author of *Cursing in America* and *Why We Curse,* has explored the dynamics of cursing in great detail. The investigators have found, among other things, that men generally curse more than women, unless said women are in a sorority, and that university provosts swear more than librarians or the staff members of the university day care center.

Regardless of who is cursing or what the provocation may be, Jay said, the rationale for the eruption is often the same. "Time and again, people have told me that cursing is a coping mechanism for them, a way of reducing stress," he said in a telephone interview. "It's a form of anger management that is often underappreciated."

Indeed, chimpanzees engage in what appears to be a kind of cursing match as a means of venting aggression and avoiding a potentially dangerous physical clash. Frans de Waal, a professor of primate behavior at Emory University in Atlanta, said that when chimpanzees are angry "they will grunt or spit or make an abrupt, upsweeping gesture that, if a human were to do it, you'd recognize it as aggressive." Such behaviors are threat gestures, Professor de Waal said, and they are all a good sign. "A chimpanzee who is really gearing up for a fight doesn't waste time with gestures, but just goes ahead and attacks," he added.

By the same token, he said, nothing is more deadly than a person who is too enraged for expletives — who cleanly and quietly picks up a gun and starts shooting.

Researchers have also examined how words attain the status of forbidden speech and how the evolution of coarse language affects the smoother sheets of civil discourse stacked above it. They have found that what counts as taboo language in a given culture is often a mirror into that culture's fears and fixations.

"In some cultures, swear words are drawn mainly from sex and bodily functions, whereas in others, they're drawn mainly from the domain of religion," Deutscher said. In societies where the purity and honor of women is of paramount importance, he said, "it's not surprising that many swear words are variations on the 'son of a whore' theme or refer graphically to the genitalia of the person's mother or sisters."

The very concept of a swear word or an oath originates from the profound importance that ancient cultures placed on swearing by the name of a god or gods. In ancient Babylon, swearing by the name of a god was meant to give absolute certainty against lying, Deutscher said, "and people believed that swearing falsely by a god would bring the terrible wrath of that god upon them." A warning against any abuse of the sacred oath is reflected in the biblical commandment that one must not "take the Lord's name in vain," and even today courtroom witnesses swear on the Bible that they are telling the whole truth and nothing but.

Among Christians, the stricture against taking the Lord's name in vain extended to casual allusions to God's son or the son's corporeal sufferings — no mention of the blood or the wounds or the body, and that goes for clever contractions, too. Nowadays the phrase "Oh, golly!" may be considered almost comically wholesome, but it was not always so. "Golly" is a compaction of "God's body" and thus was once a profanity.

Yet neither biblical commandment nor the most zealous Victorian censor can elide from the human mind its handwringing over the unruly human body, its chronic, embarrassing demands, and its sad decay. Discomfort over body functions never sleeps, Burridge said, and the need for an ever-fresh selection of euphemisms about dirty subjects has long served as an impressive engine of linguistic invention. Once a word becomes too closely associated with a specific body function, she said, once it becomes too evocative of what should not be evoked, it starts to enter the realm of the taboo and must be replaced by a new, gauzier euphemism.

For example, the word "toilet" stems from the French word for "little towel" and was originally a pleasantly indirect way of referring to the place where the chamber pot or its equivalent resides.

But toilet has since come to mean the porcelain fixture itself, and so sounds too blunt to use in polite company. Instead, you ask your tuxedoed waiter for directions to the ladies' room or the restroom or, if you must, the bathroom.

Similarly, the word "coffin" originally meant an ordinary box, but once it became associated with death, that was it for a "shoe coffin" or "thinking outside the coffin." The taboo sense of a word, Burridge said, "always drives out any other senses it might have had."

Scientists have lately sought to map the neural topography of forbidden speech by studying Tourette's patients who suffer from coprolalia, the pathological and uncontrollable urge to curse. Tourette's syndrome is a neurological disorder of unknown origin characterized predominantly by chronic motor and vocal tics, a constant grimacing or pushing of one's glasses up the bridge of one's nose or emitting a stream of small yips or grunts.

Just a small percentage of Tourette's patients have coprolalia — estimates range from 8 to 30 percent — and patient advocates are dismayed by popular portrayals of Tourette's as a humorous and invariably scatological condition. But for those who do have coprolalia, said Carlos Singer, director of the division of movement disorders at the University of Miami School of Medicine, the symptom is often the most devastating and humiliating aspect of their condition. Not only can it be shocking to people to hear a loud volley of expletives erupt for no apparent reason, sometimes from the mouth of a child or young teenager, but the curses can also be provocative and personal, florid slurs against the race, sexual identity, or body size of a passerby, for example, or deliberate and repeated lewd references to an old lover's name while in the arms of a current partner or spouse.

Reporting in the *Archives of General Psychiatry,* David A. Silbersweig, a director of neuropsychiatry and neuroimaging at the Weill Medical College of Cornell University, and his colleagues described their use of PET scans to measure cerebral blood flow and identify which regions of the brain are galvanized in Tourette's patients during episodes of tics and coprolalia. They found strong activation of the basal ganglia, a quartet of neuron clusters deep in the forebrain at roughly the level of the mid-forehead, which are known to help coordinate body movement, along with activation of crucial regions of the left rear forebrain, which participate in comprehending and generating speech, most notably Broca's area.

The researchers also saw arousal of neural circuits that interact with the limbic system, the wishbone-shaped throne of human emotions, and, significantly, of the "executive" realms of the brain, where decisions to act or desist from acting may be carried out: the neural source, scientists said, of whatever conscience, civility, or free will humans can claim.

That the brain's executive overseer is ablaze in an outburst of coprolalia, Silbersweig said, demonstrates how complex an act the urge to speak the unspeakable may be, and not only in the case of Tourette's. The person is gripped by a desire to curse, to voice something wildly inappropriate. Higher-order linguistic circuits are tapped to contrive the content of the curse. The brain's impulse control center struggles to short-circuit the collusion between limbic system urge and neocortical craft, and it may succeed for a time. Yet the urge mounts, until at last the speech pathways fire, the verboten is spoken, and the archaic and the refined brain alike must shoulder the blame.

DRAKE BENNETT

Dr. Ecstasy

FROM *The New York Times Magazine*

ALEXANDER SHULGIN, Sasha to his friends, lives with his wife, Ann, thirty minutes inland from San Francisco Bay on a hillside dotted with valley oak, Monterey pine, and hallucinogenic cactus. At seventy-nine, he stoops a little, but he is still well over six feet tall, with a mane of white hair, a matching beard, and a wardrobe that runs toward sandals, slacks, and short-sleeved shirts with vaguely ethnic patterns. He lives modestly, drawing income from a small stock portfolio supplemented by his Social Security and the rent that two phone companies pay him to put cell towers on his land. In many respects he might pass for a typical Contra Costa County retiree.

It was an acquaintance of Shulgin's named Humphry Osmond, a British psychiatrist and researcher into the effects of mescaline and LSD, who coined the word "psychedelic" in the late 1950s for a class of drugs that significantly alter one's perception of reality. Derived from Greek, the term translates as "mind manifesting" and is preferred by those who believe in the curative power of such chemicals. Skeptics tend to call them hallucinogens.

Shulgin is in the former camp. There's a story he likes to tell about the past one hundred years: "At the beginning of the twentieth century, there were only two psychedelic compounds known to Western science: cannabis and mescaline. A little over fifty years later — with LSD, psilocybin, psilocin, TMA, several compounds based on DMT and various other isomers — the number was up to almost twenty. By 2000 there were well over two hundred. So you see, the growth is exponential." When I asked him whether that meant that by 2050 we'll be up to two thousand, he smiled and said, "The way it's building up now, we may have well over that number."

The point is clear enough: the continuing explosion in options for chemical mind manifestation is as natural as the passage of time. But what Shulgin's narrative leaves out is the fact that most of this supposedly inexorable diversification took place in a lab in his backyard. For forty years, working in plain sight of the law and publishing his results, Shulgin has been a one-man psychopharmacological research sector. (Timothy Leary called him one of the century's most important scientists.) By Shulgin's own count, he has created nearly two hundred psychedelic compounds, among them stimulants, depressants, aphrodisiacs, "empathogens," convulsants, drugs that alter hearing, drugs that slow one's sense of time, drugs that speed it up, drugs that trigger violent outbursts, drugs that deaden emotion — in short, a veritable lexicon of tactile and emotional experience. And in 1976 Shulgin fished an obscure chemical called MDMA out of the depths of the chemical literature and introduced it to the wider world, where it came to be known as Ecstasy.

In the small subculture that truly believes in better living through chemistry, Shulgin's oeuvre has made him an icon and a hero: part pioneer, part holy man, part connoisseur. As his supporters point out, his work places him in an old, and in many cultures venerable, tradition. Whether it's West African *iboga* ceremonies or Navajo peyote rituals, 1960s LSD culture or the age-old cultivation of cannabis nearly everywhere on the planet it can grow, the pursuit and celebration of chemically induced alternate realms of consciousness goes back beyond the dawn of recorded history and has proved impossible to fully suppress. Shulgin sees nothing strange about devoting his life to it. What's strange to him is that so few others see fit to do the same thing.

Most of the scientific community considers Shulgin at best a curiosity and at worst a menace. Now, however, near the end of his career, his faith in the potential of psychedelics has at least a chance at vindication. In December 2004 the Food and Drug Administration approved a Harvard Medical School study looking at whether MDMA can alleviate the fear and anxiety of terminal cancer patients. And February 2005 marked a year since Michael Mithoefer, a psychiatrist in Charleston, South Carolina, started his study of Ecstasy-assisted therapy for posttraumatic stress disorder. At the same time, with somewhat less attention, studies at the Harbor-UCLA Medical Center and the University of Arizona, Tucson, have focused on the therapeutic potential of psilocybin (the active ingredient in "magic mushrooms"). It's far from a revolution, but it is an opening, and as both scientist and advocate, Shulgin has helped create it. If — and it's a big if — the results of the studies are promising enough, it might bring something like legitimacy to the Shulgin pharmacopoeia.

"I've always been interested in the machinery of the mental process," Shulgin told me not long ago. He has also, from a very young age, loved playing with chemicals. As a lonely sixteen-year-old Harvard scholarship student soon to drop out and join the navy, he studied organic chemistry. His interest in pharmacology dates to 1944, when a military nurse gave him some orange juice just before his surgery for a thumb infection. Convinced that the undissolved crystals at the bottom of the glass were a sedative, Shulgin fell unconscious, only to find upon waking that the substance had been sugar. It was a revelatory, tantalizing hint of the mind's odd strength.

When Shulgin had his first psychedelic experience in 1960, he was a young UC Berkeley biochemistry Ph.D. working at Dow Chemical. He had been interested for several years in the chemistry of mescaline, the active ingredient in peyote, and one spring day a few friends offered to keep an eye on him while he tried it himself. He spent the afternoon enraptured by his surroundings. Most important, he later wrote, he realized that everything he saw and thought "had been brought about by a fraction of a gram of a white solid, but that in no way whatsoever could it be argued that these memories had been contained within the white solid . . . I understood that our entire universe is contained in the mind and the spirit. We may choose not to find access to it, we may even deny its existence, but it is indeed there inside us, and there are chemicals that can catalyze its availability."

Epiphanies don't come much grander than that, and Shulgin's interest in psychoactive drugs bloomed into an obsession. "There was," he remembers thinking, "this remarkably rich and unexplored area that I had to explore." Two years later he was given his chance when he created Zectran, one of the world's first biodegradable insecticides. In return, Dow gave him its customary dollar for the patent and unlimited freedom to pursue his interests.

As Shulgin turned toward making psychedelics, Dow remained true to its word. When the company asked, he patented his compounds. When it didn't, Shulgin published his findings in places like *Nature* and the *Journal of Organic Chemistry*. Eventually, however, Dow decided that Shulgin's work wasn't something it wanted to endorse and asked that he not use the company address in his publications. He began to work out of a lab he had set up at home, eventually leaving Dow altogether to freelance as a consultant to research labs and hospitals.

All along he made drugs: 2,5-dimethoxy-4-ethoxyamphetamine, or MEM for short, was his Rosetta stone, a "valuable and dramatic compound" that opened the door to a whole class of drugs based on changes at the "4 position" of a molecule's central carbon ring. A compound he dubbed Aleph-1 gave him "one of the most delicious blends of inflation, paranoia, and selfishness that I have ever experienced." Another, Ariadne, was patented and tested under the name Dimoxamine as a drug for "restoring motivation in senile geriatric patients." Still another, DIPT, created no visual hallucinations but distorted the user's sense of pitch.

Shulgin tested the chemicals' activity by taking them himself. He would start many times below the active dose of a compound's closest analogue and work his way up on alternate days. When he found something of interest, Ann, whom he married in 1981, would try it. If he thought further study was warranted, he would invite over his "research group" of six to eight close friends — among them two psychologists and a fellow chemist — and try the drugs out on them. In case of a truly dangerous reaction, Shulgin kept an anticonvulsant on hand. He used it twice, both times on himself.

Shulgin's pace has slowed recently — the research group hardly meets anymore. Nevertheless, Ann figures that she's had more than two thousand psychedelic experiences. Shulgin puts his own figure above four thousand. Asked if they had suffered any effects from their remarkable drug histories, they laughed. "You mean negative effects?" Ann said. In more than a dozen hours of conversation, her memory proved sharp. But Shulgin, while a nimble conversationalist, can have trouble with names — of people and places but never of chemicals. At one point, while explaining a mnemonic device he uses to remember world geography, he paused and asked me, "Where's that place where Ann is from?" (She was born in New Zealand.) He is, though, also nearing eighty.

Once a Shulgin compound develops a reputation, it is almost invariably placed on the Drug Enforcement Agency's list of Schedule I drugs, those deemed to have no accepted medical use and the highest potential for abuse or addiction. It is therefore rather striking that Shulgin is not only still a free man but also still at work. His own explanation is that, quite simply, "I'm not doing anything illegal." For more than twenty years, until a government crackdown, he had a DEA-issued Schedule I research license. And many of the drugs in his lab weren't illegal because they hadn't existed until he created them.

Shulgin's knack for befriending the right people hasn't hurt. A week after I visited him, he was headed to Sonoma County for the annual "summer encampment" of the Bohemian Club, an exclusive, secretive San Francisco–based men's club that has counted every Republican president since Herbert Hoover among its members.

For a long time, though, Shulgin's most helpful relationship was with the DEA itself. The head of the DEA's Western Laboratory, Bob Sager, was one of his closest friends. Sager officiated at the Shulgins' wedding and, a year later, was married on Shulgin's lawn. Through Sager, the agency came to rely on Shulgin: he would give pharmacology talks to the agents, make drug samples for the forensic teams, and serve as an expert witness — though, he is quick to point out, he appeared much more frequently for the defense. He even wrote the definitive law-enforcement desk-reference work on controlled substances. In his office, Shulgin has several plaques awarded to him by the agency for his service. (Shulgin denies that this had anything to do with his being given his Schedule I license.)

Nevertheless, in the early 1980s, Shulgin began having grim fantasies of the DEA throwing him in jail, ransacking his lab, and destroying all of his records. At the same time, he was finding it harder to get his work published: journals were either uninterested in or leery about human psychedelic research. He decided to make as much of what he knew public as quickly as possible. He and Ann started work on a book called *PiHKAL* (short for "Phenethylamines I Have Known and Loved," after a family of compounds particularly rich in psychoactivity), self-publishing it in 1991.

It is a curious hybrid work, divided into two sections. The first, "The Love Story," is a thinly fictionalized account of Sasha's and Ann's comings of age, previous marriages, meeting, courtship (to which nearly two hundred pages are devoted), and many drug experiences. The second, "The Chemical Story," is not a story at all but capsule descriptions of 179 phenethylamines. Each entry includes step-by-step instructions for synthesis, along with recommended dosages, duration of action, and "qualitative comments" like the following, for 60 milligrams of something called 3C-E: "Visuals very strong, insistent. Body discomfort remained very heavy for first hour . . . 2nd hour on, bright colors, distinct shapes — jewel-like — with eyes closed. Suddenly it became clearly not anti-erotic . . . Image of glass-walled apartment building in mid-desert. Exquisite sensitivity. Down by? midnight. Next morning, faint flickering lights on looking out windows." *TiHKAL* ("Tryptamines I Have Known and Loved"), self-published six years later, follows the same model.

To date, *PiHKAL* has sold more than forty-one thousand copies, a figure nearly unheard-of for a self-published book. It introduced Shulgin's work to a whole new audience and turned him into an underground celebrity. An organization called the Center for Cognitive Liberty and Ethics has an online *Ask Dr. Shulgin* column that receives two hundred questions a month. On independent drug-information Web sites like www.erowid.com, you can find the *PiHKAL* and *TiHKAL* entries for dozens of drugs, along with many anonymously posted accounts of Shulgin-style self-dosing drug experiments, some of them harrowing in their recklessness.

With all of these fellow travelers, some very bad experiences are inevitable. In 1967 a Shulgin compound called DOM enjoyed a brief vogue in Haight-Ashbury under the name STP, at doses several times larger than those at which Shulgin had found significant psychoactive effects, and emergency rooms saw a spike in the number of people coming in thinking they would never come down. And while the number of psychedelic-related deaths is orders of magnitude smaller than the number due to alcohol, prescription drugs, or even over-the-counter painkillers, they do occur regularly. In October 2000 a twenty-year-old man in Norman, Oklahoma, died from taking 2C-T-7, a drug Shulgin describes in *PiHKAL* as "good and friendly and wonderful."

When I asked Shulgin whether he remembered the first time he heard that someone had died from one of his drugs, he said he did not: "It would have struck me as being a sad event. And yet, at the same time, how many people die from aspirin? It's a small but real percentage." (The American Association of Poison Control Centers, whose numbers are not comprehensive, attributed fifty-nine deaths to aspirin in 2003; most, though, were suicides.) Asked whether he could imagine a drug so addictive that it should be banned, he said no. With his fervent libertarianism — he says the only appropriate restriction on drugs is one to prevent children from buying them — he has inoculated himself against any sense of personal guilt.

Shulgin's special relationship with the DEA ended two years after the publication of *PiHKAL*. According to Richard Meyer, spokesman for the agency's San Francisco Field Division: "It is our opinion that those books are pretty much cookbooks on how to make illegal drugs. Agents tell me that in clandestine labs that they have raided, they have found copies of those books." In 1993 DEA agents descended on Shulgin's farm, combed through the house and lab, and carted off anything they thought might be an illicit substance. Shulgin was fined $25,000 for violations of the terms of his Schedule I license (donations from friends and admirers ended up covering the whole amount) and was asked to turn the license in.

To the extent that Shulgin is known to the wider world, it is as the godfather of Ecstasy: 3,4-methylenedioxy-N-methylamphetamine, or MDMA, was originally patented in 1914 by Merck. The byproduct of a chemical synthesis, it was thought to have no use of its own and was promptly forgotten. But Shulgin resynthesized it in 1976 at the suggestion of a former student. (He has never found out how she heard about it.) Two years later, in a paper written with his friend and fellow chemist David Nichols, he was the first to publicly document its effect on humans: "an easily controlled altered state of consciousness with emotional and sensual overtones."

Unlike many of its subsequent users, Shulgin did not find his

MDMA experience transformative. For him the effect was like a particularly lucid alcohol buzz; he called it his "low-calorie martini." He was intrigued, though, by the drug's unique combination of intoxication, disinhibition, and clarity. "It didn't have the other visual and auditory imaginative things that you often get from psychedelics," he said. "It opened up a person, both to other people and inner thoughts, but didn't necessarily color it with pretty colors and strange noises." He decided that it might be well suited for psychotherapy.

At the time it was not such an unconventional idea. In the fifties and sixties, the use of LSD, psilocybin, and mescaline in therapy was the subject of much mainstream scholarly debate. LSD was a particularly hot topic: more than a thousand papers were written on its use as an experimental treatment for alcoholism, depression, and various neuroses in some forty thousand patients. One proponent was a psychotherapist and friend of Shulgin's named Leo Zeff. When Shulgin had him try MDMA in 1977, Zeff was so impressed that he came out of retirement to proselytize for it. Ann Shulgin remembers a speaker at Zeff's memorial service saying that Zeff had introduced the drug to "about four thousand" therapists.

In certain therapeutic circles, MDMA acquired a reputation as a wonder drug. Anecdotal accounts attested to its ability to induce in one session the sort of breakthroughs that normally took months or years of therapy. According to George Greer, a psychiatrist who in the early eighties conducted MDMA therapy sessions with eighty patients, "Without exception, every therapist who I talked to or even heard of, every therapist who gave MDMA to a patient, was highly impressed by the results."

But the drug was also showing up in nightclubs in Dallas and Los Angeles, and in 1986 the DEA placed it in Schedule I. By the late nineties, household surveys showed millions of teenagers and college students using it, and in 2000, U.S. Customs officials seized nearly ten million pills. Parents and public officials worried that a whole generation was consigning itself to a life of drug-induced depression and cognitive decay.

There is, in fact, little consensus about what MDMA does to your brain over the long run. Researchers generally agree on its immediate physiological effects: especially at higher doses, it can trigger sharp increases in muscle tension, heart rate, and blood pressure. Hyperthermia, or raised body temperature, is a particular worry, along with the attendant risk of heat stroke or dehydration. MDMA also, at least temporarily, exhausts the brain's supply of serotonin (a neurochemical thought to play a role in memory and mood regulation). But as to the extent and duration of that depletion, and whether it has any measurable functional or behavioral consequences, there is fierce debate and surprisingly scarce data. Nationwide, fatality numbers are hard to come by, but a study by New York City's deputy chief medical examiner determined that of the 19,000 deaths from all causes reported to his office between January 1997 and June 2000, 2 were due solely to Ecstasy.

In the past couple of years, MDMA's opponents have backed off from some of their stronger claims. (In one particularly embarrassing instance, a study linking MDMA to Parkinson's disease was revealed to have instead been based on the use of methamphetamine, which is known to be much more neurotoxic.) Emboldened, a few psychiatrists are bringing MDMA back into the news in a role closer to the one Shulgin originally imagined for it.

With the FDA's approval of the Harvard cancer-patient study on December 17, 2004, all that's still needed is a DEA license for MDMA. John Halpern, the psychiatrist heading the study, anticipates that happening in the next couple of months. At the same time, he cautions against making too much of his "small pilot study": eight subjects undergoing a course of MDMA therapy, with another four receiving a placebo. The Charleston study is similarly modest, with twenty subjects.

Still, according to Mark A. R. Kleiman, director of the Drug Policy Analysis Program at UCLA, "there's obviously been a significant shift at the regulatory agencies and the Institutional Review Boards. There are studies being approved that wouldn't have been approved ten years ago. And there are studies being proposed that wouldn't have been proposed ten years ago."

The theoretical basis for MDMA therapy varies a bit depending on whom you talk to. Greer says that by lowering patients' defenses, the drug allows them to face troubling, even repressed, memories. Charles Grob, the psychiatry professor running the UCLA psilocybin study (also with terminal cancer patients) and a longtime advocate of therapeutic MDMA research, focuses more on the "empathic rapport" catalyzed by MDMA. "I don't know of any other compound that can achieve this to the degree that MDMA can," he said.

The medical community remains dubious. For Vivian Rakoff, emeritus professor of psychiatry at the University of Toronto, there is something familiar about the claims being made for psychedelics. "The notion of the revelatory moment due to some drug or maneuver that will allow you to change your life has been around for a long time," he said. "Every few years, something comes along that claims to be what Freud called the 'royal road to the unconscious.'" Steven Hyman, professor of neurobiology at Harvard Medical School and former director of the National Institute of Mental Health, put it this way: "If you asked me to place a bet, I would be skeptical. In general, one worries that insights gained under states of disinhibition or mild euphoria or different cognitive states with illusions may seem strange and distant from the vantage of our ordinary life." Even so, both Hyman and Rakoff say that research should be allowed to proceed.

Shulgin has been credited with jump-starting today's therapeutic research, but he prefers to play down his role. While heartened by the MDMA studies and happy to play psychedelic elder statesman, he insists that he is not a healer or a shaman but a researcher. Asked why he does what he does, he replies, "I'm curious!" He is most animated when describing the feeling that accompanies the discovery of a new compound, no matter what its properties. Sometimes he compares the moment to that of artistic creation ("The pleasure of composing a new painting or piece of music"), and sometimes it sounds more like a close encounter of the third kind ("You're meeting something you don't know, and it's meeting something it doesn't know. And so you have this exchange of properties and ideas").

Shulgin's lab is in the concrete-block foundation of what used to be a small cabin, set into a ridge a few dozen yards from his house along a narrow brick path. On the door is a laminated sign that reads, "This is a research facility that is known to and authorized by the Contra Costa County Sheriff's Office, all San Francisco DEA Personnel and the State and Federal EPA Authorities." Underneath are phone numbers for the relevant official at each agency.

He posted it after the sheriff's department and the DEA raided the farm a second time a few years ago. (They later apologized.)

Shulgin gave me my tour late one afternoon. A weak light came in through the small, dusty windows. The smell — synthetic and organic at once, like a burning tire doused in urine — took some getting used to. Bulbous flasks were clipped into place above a counter crowded with glassware shaped like finds from the Burgess Shale. "Everything you need is right here," Shulgin declared, pulling out drawer after clattering drawer of test tubes, beakers, plastic tubing, and syringes. At the far end of the room, beside the fireplace, was a small chalkboard covered with the traces of his brainstorming — antennaed pentagons and hexagons ringed with N's, H's, C's, and O's. Shulgin picked a short bit of scrap wood off the counter. He occasionally used it, he explained, to tear down the spider webs that festooned the rafters. "But the main problem is the squirrels," he said, pointing to where he had put up sheet metal to keep them out. "It doesn't look like the labs you see in the movies, but you get a chemist out here, and he'll say, 'Oh, my God, I'd love to have a lab like this.'"

Of course, in a way, it's exactly the sort of lab that you see in the movies — they're just movies in which the scientists wear frock coats, turn into monsters, and abduct wan women in nightgowns. There's an undeniable romance to what Shulgin does. As he stood there with his spider-web stick, describing what it's like to be in the lab late on a cold night with the fire blazing and Rachmaninoff on the radio, it seemed to me that he realized it.

He might best be described not as a scientist in the modern sense but as a different type — what Aldous Huxley, the novelist turned psychedelic philosopher, once described as a "naturalist of the mind," a "collector of psychological specimens" whose "primary concern was to make a census, to catch, kill, stuff, and describe as many kinds of beasts as he could lay his hands on." Shulgin has on occasion run PET scans to see where in the brain some of his drugs go. He has offered theories as to mechanisms of action or, as with MDMA, even suggested an application for a drug. But his primary purpose, as he sees it, is not to worry about things like that — much less about the political and social consequences of his creations. His job is to be first and then push on somewhere new. What to do with the widening wake of chemicals he leaves behind is for the rest of us to figure out.

LARRY CAHILL

His Brain, Her Brain

FROM *Scientific American*

ON A GRAY DAY in mid-January, Lawrence Summers, the president of Harvard University, suggested that innate differences in the build of the male and female brain might be one factor underlying the relative scarcity of women in science. His remarks reignited a debate that has been smoldering for a century, ever since some scientists sizing up the brains of both sexes began using their main finding — that female brains tend to be smaller — to bolster the view that women are intellectually inferior to men.

To date, no one has uncovered any evidence that anatomical disparities might render women incapable of achieving academic distinction in math, physics, or engineering. And the brains of men and women have been shown to be quite clearly similar in many ways. Nevertheless, over the past decade investigators have documented an astonishing array of structural, chemical, and functional variations between the brains of males and females.

These inequities are not just interesting idiosyncrasies that might explain why more men than women enjoy the Three Stooges. They raise the possibility that we might need to develop sex-specific treatments for a host of conditions, including depression, addiction, schizophrenia, and posttraumatic stress disorder (PTSD). Furthermore, the differences imply that researchers exploring the structure and function of the brain must take into account the sex of their subjects when analyzing their data — and include both women and men in future studies or risk obtaining misleading results.

Sculpting the Brain

Not so long ago neuroscientists believed that sex differences in the brain were limited mainly to those regions responsible for mating behavior. In a 1966 *Scientific American* article titled "Sex Differences in the Brain," Seymour Levine of Stanford University described how sex hormones help to direct divergent reproductive behaviors in rats — with males engaging in mounting and females arching their backs and raising their rumps to attract suitors. Levine mentioned only one brain region in his review: the hypothalamus, a small structure at the base of the brain that is involved in regulating hormone production and controlling basic behaviors such as eating, drinking, and sex. A generation of neuroscientists came to maturity believing that "sex differences in the brain" referred primarily to mating behaviors, sex hormones, and the hypothalamus.

That view, however, has now been knocked aside by a surge of findings that highlight the influence of sex on many areas of cognition and behavior, including memory, emotion, vision, hearing, the processing of faces, and the brain's response to stress hormones. This progress has been accelerated in the past five to ten years by the growing use of sophisticated noninvasive imaging techniques such as positron emission tomography (PET) and functional magnetic resonance imaging (fMRI), which can peer into the brains of living subjects.

These imaging experiments reveal that anatomical variations occur in an assortment of regions throughout the brain. Jill M. Goldstein of Harvard Medical School and her colleagues, for example, used MRI to measure the sizes of many cortical and subcortical areas. Among other things, these investigators found that parts of the frontal cortex, the seat of many higher cognitive functions, are bulkier in women than in men, as are parts of the limbic cortex, which is involved in emotional responses. In men, on the other hand, parts of the parietal cortex, which is involved in space perception, are bigger than in women, as is the amygdala, an almond-shaped structure that responds to emotionally arousing information — to anything that gets the heart pumping and the adrenaline flowing. These size differences, as well as others mentioned throughout the article, are relative: they refer to the overall volume of the structure relative to the overall volume of the brain.

Differences in sizes of brain structures are generally thought to reflect their relative importance to the animal. For example, primates rely more on vision than olfaction; for rats, the opposite is true. As a result, primate brains maintain proportionately larger regions devoted to vision, and rat brains devote more space to olfaction. So the existence of widespread anatomical disparities between men and women suggests that sex does influence the way the brain works.

Other investigations are finding anatomical sex differences at the cellular level. For example, Sandra Witelson and her colleagues at McMaster University discovered that women possess a greater density of neurons in parts of the temporal lobe cortex associated with language processing and comprehension. On counting the neurons in postmortem samples, the researchers found that of the six layers present in the cortex, two show more neurons per unit volume in females than in males. Similar findings were subsequently reported for the frontal lobe. With such information in hand, neuroscientists can now explore whether sex differences in neuron number correlate with differences in cognitive abilities — examining, for example, whether the boost in density in the female auditory cortex relates to women's enhanced performance on tests of verbal fluency.

Such anatomical diversity may be caused in large part by the activity of the sex hormones that bathe the fetal brain. These steroids help to direct the organization and wiring of the brain during development and influence the structure and neuronal density of various regions. Interestingly, the brain areas that Goldstein found to differ between men and women are ones that in animals contain the highest number of sex hormone receptors during development. This correlation between brain region size in adults and sex steroid action in utero suggests that at least some sex differences in cognitive function do not result from cultural influences or the hormonal changes associated with puberty — they are there from birth.

Inborn Inclinations

Several intriguing behavioral studies add to the evidence that some sex differences in the brain arise before a baby draws its first breath. Through the years, many researchers have demonstrated that when selecting toys, young boys and girls part ways. Boys tend to gravitate toward balls or toy cars, whereas girls more typically reach for a doll. But no one could really say whether those preferences are dictated by culture or by innate brain biology.

To address this question, Melissa Hines of City University London and Gerianne M. Alexander of Texas A&M University turned to monkeys, one of our closest animal cousins. The researchers presented a group of vervet monkeys with a selection of toys, including rag dolls, trucks, and some gender-neutral items such as picture books. They found that male monkeys spent more time playing with the "masculine" toys than their female counterparts did, and female monkeys spent more time interacting with the playthings typically preferred by girls. Both sexes spent equal time monkeying with the picture books and other gender-neutral toys.

Because vervet monkeys are unlikely to be swayed by the social pressures of human culture, the results imply that toy preferences in children result at least in part from innate biological differences. This divergence, and indeed all the anatomical sex differences in the brain, presumably arose as a result of selective pressures during evolution. In the case of the toy study, males — both human and primate — prefer toys that can be propelled through space and that promote rough-and-tumble play. These qualities, it seems reasonable to speculate, might relate to the behaviors useful for hunting and for securing a mate. Similarly, one might also hypothesize that females select toys that allow them to hone the skills they will one day need to nurture their young.

Simon Baron-Cohen and his associates at the University of Cambridge took a different but equally creative approach to addressing the influence of nature versus nurture regarding sex differences. Many researchers have described disparities in how "people-centered" male and female infants are. For example, Baron-Cohen and his student Svetlana Lutchmaya found that one-year-old girls spend more time looking at their mothers than boys of the same age do. And when these babies are presented with a choice of films to watch, the girls look longer at a film of a face, whereas boys lean toward a film featuring cars.

Of course, these preferences might be attributable to differences in the way adults handle or play with boys and girls. To eliminate this possibility, Baron-Cohen and his students went a step further. They took their video camera to a maternity ward to examine the preferences of babies that were only one day old. The infants saw either the friendly face of a live female student or a mobile that matched the color, size, and shape of the student's face and included a scrambled mix of her facial features. To avoid any bias, the experimenters were unaware of each baby's sex during testing. When they watched the tapes, they found that the girls spent more time looking at the student, whereas the boys spent more time looking at the mechanical object. This difference in social interest was evident on day one of life — implying again that we come out of the womb with some cognitive sex differences built in.

Under Stress

In many cases, sex differences in the brain's chemistry and construction influence how males and females respond to the environment or react to, and remember, stressful events. Take, for example, the amygdala. Goldstein and others have reported that the amygdala is larger in men than in women. And in rats, the neurons in this region make more numerous interconnections in males than in females. These anatomical variations would be expected to produce differences in the way that males and females react to stress.

To assess whether male and female amygdalae in fact respond differently to stress, Katharina Braun and her coworkers at Otto von Guericke University in Magdeburg, Germany, briefly removed a litter of degu pups from their mother. For these social South American rodents, which live in large colonies like prairie dogs, even temporary separation can be quite upsetting. The researchers then measured the concentration of serotonin receptors in various brain regions. Serotonin is a neurotransmitter, or signal-carrying molecule, that is key for mediating emotional behavior. (Prozac, for example, acts by increasing serotonin function.)

The workers allowed the pups to hear their mother's call during the period of separation and found that this auditory input increased the serotonin-receptor concentration in the males' amygdala yet decreased the concentration of these same receptors in females. Although it is difficult to extrapolate from this study to human behavior, the results hint that if something similar occurs in children, separation anxiety might differentially affect the emotional well-being of male and female infants. Experiments such as these are necessary if we are to understand why, for instance, anxiety disorders are far more prevalent in girls than in boys.

Another brain region now known to diverge in the sexes anatomically and in its response to stress is the hippocampus, a structure crucial for memory storage and for spatial mapping of the physical environment. Imaging consistently demonstrates that the hippocampus is larger in women than in men. These anatomical differences may well relate somehow to differences in the way males and females navigate. Many studies suggest that men are more likely to navigate by estimating distance in space and orientation ("dead reckoning"), whereas women are more likely to navigate by monitoring landmarks. Interestingly, a similar sex difference exists in rats. Male rats are more likely to navigate mazes using directional and positional information, whereas female rats are more likely to navigate the same mazes using available landmarks. (Investigators have yet to demonstrate, however, that male rats are less likely to ask for directions.)

Even the neurons in the hippocampus behave differently in males and females, at least in how they react to learning experiences. For example, Janice M. Juraska and her associates at the University of Illinois have shown that placing rats in an "enriched environment" — cages filled with toys and with fellow rodents to promote social interactions — produced dissimilar effects on the structure of hippocampal neurons in male and female rats. In females, the experience enhanced the "bushiness" of the branches in the cells' dendritic trees — the many-armed structures that receive signals from other nerve cells. This change presumably reflects an increase in neuronal connections, which in turn is thought to be involved with the laying down of memories. In males, however, the complex environment either had no effect on the dendritic trees or pruned them slightly.

But male rats sometimes learn better in the face of stress. Tracey J. Shors of Rutgers University and her collaborators have found that a brief exposure to a series of one-second tail shocks enhanced performance of a learned task and increased the density of dendritic connections to other neurons in male rats yet impaired performance and decreased connection density in female rats. Findings such as these have interesting social implications. The more we discover about how brain mechanisms of learning differ between the sexes, the more we may need to consider how optimal learning environments potentially differ for boys and girls.

Although the hippocampus of the female rat can show a decrement in response to acute stress, it appears to be more resilient than its male counterpart in the face of chronic stress. Cheryl D. Conrad and her coworkers at Arizona State University restrained rats in a mesh cage for six hours — a situation that the rodents find disturbing. The researchers then assessed how vulnerable their hippocampal neurons were to killing by a neurotoxin — a standard measure of the effect of stress on these cells. They noted that chronic restraint rendered the males' hippocampal cells more susceptible to the toxin but had no effect on the females' vulnerability. These findings, and others like them, suggest that in terms of brain damage, females may be better equipped to tolerate chronic stress than males are. Still unclear is what protects female hippocampal cells from the damaging effects of chronic stress, but sex hormones very likely play a role.

The Big Picture

Extending the work on how the brain handles and remembers stressful events, my colleagues and I have found contrasts in the way men and women lay down memories of emotionally arousing incidents — a process known from animal research to involve activation of the amygdala. In one of our first experiments with human subjects, we showed volunteers a series of graphically violent films while we measured their brain activity using PET. A few weeks later we gave them a quiz to see what they remembered.

We discovered that the number of disturbing films they could recall correlated with how active their amygdala had been during the viewing. Subsequent work from our laboratory and others confirmed this general finding. But then I noticed something strange. The amygdala activation in some studies involved only the right hemisphere, and in others it involved only the left hemisphere. It was then I realized that the experiments in which the right amygdala lit up involved only men; those in which the left amygdala was fired up involved women. Since then, three subsequent studies — two from our group and one from John Gabrieli and Turhan Canli and their collaborators at Stanford — have confirmed this difference in how the brains of men and women handle emotional memories.

The realization that male and female brains were processing the same emotionally arousing material into memory differently led us to wonder what this disparity might mean. To address this question, we turned to a century-old theory stating that the right hemisphere is biased toward processing the central aspects of a situation, whereas the left hemisphere tends to process the finer details. If that conception is true, we reasoned, a drug that dampens the activity of the amygdala should impair a man's ability to recall the gist of an emotional story (by hampering the right amygdala) but should hinder a woman's ability to come up with the precise details (by hampering the left amygdala).

Propranolol is such a drug. This so-called beta blocker quiets the activity of adrenaline and its cousin noradrenaline and, in so doing, dampens the activation of the amygdala and weakens recall of emotionally arousing memories. We gave this drug to men and women before they viewed a short slide show about a young boy caught in a terrible accident while walking with his mother. One week later we tested their memory. The results showed that propranolol made it harder for men to remember the more holistic aspects, or gist, of the story — that the boy had been run over by a car, for example. In women, propranolol did the converse, impairing their memory for peripheral details — that the boy had been carrying a soccer ball.

In more recent investigations, we found that we can detect a hemispheric difference between the sexes in response to emotional material almost immediately. Volunteers shown emotionally unpleasant photographs react within 300 milliseconds — a response that shows up as a spike on a recording of the brain's electrical activity. With Antonella Gasbarri and others at the University of L'Aquila in Italy, we have found that in men this quick spike, termed a P300 response, is more exaggerated when recorded over the right hemisphere; in women, it is larger when recorded over the left. Hence, sex-related hemispheric disparities in how the brain processes emotional images begin within 300 milliseconds —

long before a person has had much, if any, chance to consciously interpret what he or she has seen.

These discoveries might have ramifications for the treatment of PTSD. Previous research by Gustav Schelling and his associates at Ludwig Maximilian University in Germany had established that drugs such as propranolol diminish memory of traumatic situations when administered as part of the usual therapies in an intensive care unit. Prompted by our findings, they found that, at least in such units, beta blockers reduce memory for traumatic events in women but not in men. Even in intensive care, then, physicians may need to consider the sex of their patients when meting out their medications.

Sex and Mental Disorders

PTSD is not the only psychological disturbance that appears to play out differently in women and men. A PET study by Mirko Diksic and his colleagues at McGill University showed that serotonin production was a remarkable 52 percent higher on average in men than in women, which might help clarify why women are more prone to depression — a disorder commonly treated with drugs that boost the concentration of serotonin.

A similar situation might prevail in addiction. In this case, the neurotransmitter in question is dopamine — a chemical involved in the feelings of pleasure associated with drugs of abuse. Studying rats, Jill B. Becker and her fellow investigators at the University of Michigan at Ann Arbor discovered that in females, estrogen boosted the release of dopamine in brain regions important for regulating drug-seeking behavior. Furthermore, the hormone had long-lasting effects, making the female rats more likely to pursue cocaine weeks after last receiving the drug. Such differences in susceptibility — particularly to stimulants such as cocaine and amphetamine — could explain why women might be more vulnerable to the effects of these drugs and why they tend to progress more rapidly from initial use to dependence than men do.

Certain brain abnormalities underlying schizophrenia appear to differ in men and women as well. Ruben Gur, Raquel Gur, and their colleagues at the University of Pennsylvania have spent years investigating sex-related differences in brain anatomy and function. In one project, they measured the size of the orbitofrontal cortex, a region involved in regulating emotions, and compared it with the size of the amygdala, implicated more in producing emotional reactions. The investigators found that women possess a significantly larger orbitofrontal-to-amygdala ratio (OAR) than men do. One can speculate from these findings that women might on average prove more capable of controlling their emotional reactions.

In additional experiments, the researchers discovered that this balance appears to be altered in schizophrenia, though not identically for men and women. Women with schizophrenia have a decreased OAR relative to their healthy peers, as might be expected. But men, oddly, have an increased OAR relative to healthy men. These findings remain puzzling, but, at the least, they imply that schizophrenia is a somewhat different disease in men and women and that treatment of the disorder might need to be tailored to the sex of the patient.

Sex Matters

In a comprehensive 2001 report on sex differences in human health, the prestigious National Academy of Sciences asserted that "sex matters. Sex, that is, being male or female, is an important basic human variable that should be considered when designing and analyzing studies in all areas and at all levels of biomedical and health-related research."

Neuroscientists are still far from putting all the pieces together — identifying all the sex-related variations in the brain and pinpointing their influences on cognition and propensity for brain-related disorders. Nevertheless, the research conducted to date certainly demonstrates that differences extend far beyond the hypothalamus and mating behavior. Researchers and clinicians are not always clear on the best way to go forward in deciphering the full influences of sex on the brain, behavior, and responses to medications. But growing numbers now agree that going back to assuming we can evaluate one sex and learn equally about both is no longer an option.

MICHAEL CHOROST

My Bionic Quest for Boléro

FROM *Wired*

WITH ONE LISTEN, I was hooked. I was a fifteen-year-old suburban New Jersey nerd, racked with teenage lust but too timid to ask for a date. When I came across *Boléro* among the LPs in my parents' record collection, I put it on the turntable. It hit me like a neural thunderstorm, titanic and glorious, each cycle building to a climax and waiting but a beat before launching into the next.

I had no idea back then of *Boléro*'s reputation as one of the most famous orchestral pieces in the world. When it was first performed at the Paris Opera in 1928, the fifteen-minute composition stunned the audience. Of the French composer, Maurice Ravel, a woman in attendance reportedly cried out, "He's mad . . . he's mad!" One critic wrote that *Boléro* "departs from a thousand years of tradition."

I sat in my living room alone, listening. *Boléro* starts simply enough, a single flute accompanied by a snare drum: *da-da-da-dum, da-da-da-dum, dum-dum, da-da-da-dum.* The same musical clause repeats seventeen more times, each cycle adding instruments, growing louder and more insistent, until the entire orchestra roars in an overpowering finale of rhythm and sound. Musically, it was perfect for my ear. It had a structure that I could easily grasp and enough variation to hold my interest.

It took a lot to hold my interest; I was nearly deaf at the time. In 1964 my mother contracted rubella while pregnant with me. Hearing aids allowed me to understand speech well enough, but most music was lost on me. *Boléro* was one of the few pieces I actually enjoyed. A few years later I bought the CD and played it so much it eventually grew pitted and scratched. It became my touchstone. Every time I tried out a new hearing aid, I'd check to see if *Boléro* sounded okay. If it didn't, the hearing aid went back.

And then, on July 7, 2001, at 10:30 A.M., I lost my ability to hear *Boléro* — and everything else. While I was waiting to pick up a rental car in Reno, I suddenly thought the battery in my hearing aid had died. I replaced it. No luck. I switched hearing aids. Nothing.

I got into my rental car and drove to the nearest emergency room. For reasons that are still unknown, my only functioning ear had suffered "sudden-onset deafness." I was reeling, trying to navigate in a world where the volume had been turned down to zero.

But there was a solution, a surgeon at Stanford Hospital told me a week later, speaking slowly so I could read his lips. I could have a computer surgically installed in my skull. A cochlear implant, as it is known, would trigger my auditory nerves with sixteen electrodes that snaked inside my inner ear. It seemed drastic, and the $50,000 price tag was a dozen times more expensive than a high-end hearing aid. I went home and cried. Then I said yes.

For the next two months, while awaiting surgery, I was totally deaf except for a thin trickle of sound from my right ear. I had long since become accustomed to not hearing my own voice when I spoke. It happened whenever I removed my hearing aid. But that sensation was as temporary as waking up without my glasses. Now, suddenly, the silence wasn't optional. At my job as a technical writer in Silicon Valley, I struggled at meetings. Using the phone was out of the question.

In early September, the surgeon drilled a tunnel through an inch and a half of bone behind my left ear and inserted the sixteen electrodes along the auditory nerve fibers in my cochlea. He hollowed a well in my skull about the size of three stacked quarters and snapped in the implant.

When the device was turned on a month after surgery, the first sentence I heard sounded like "Zzzzzz szz szvizzz ur brfzzzzzz?" My brain gradually learned how to interpret the alien signal. Before long, "Zzzzzz szz szvizzz ur brfzzzzzz?" became "What did you have for breakfast?" After months of practice, I could use the telephone again, even converse in loud bars and cafeterias. In many ways, my hearing was better than it had ever been. Except when I listened to music.

I could hear the drums of *Boléro* just fine. But the other instruments were flat and dull. The flutes and soprano saxophones sounded as though someone had clapped pillows over them. The oboes and violins had become groans. It was like walking colorblind through a Paul Klee exhibit. I played *Boléro* again and again, hoping that practice would bring it, too, back to life. It didn't.

The implant was embedded in my head; it wasn't some flawed hearing aid I could just send back. But it *was* a computer. Which meant that, at least in theory, its effectiveness was limited only by the ingenuity of software engineers. As researchers learn more about how the ear works, they continually revise cochlear implant software. Users await new releases with all the anticipation of Apple zealots lining up for the latest Mac OS.

About a year after I received the implant, I asked one implant engineer how much of the device's hardware capacity was being used. "Five percent maybe." He shrugged. "Ten, tops."

I was determined to use that other 90 percent. I set out on a crusade to explore the edges of auditory science. For two years I tugged on the sleeves of scientists and engineers around the country, offering myself as a guinea pig for their experiments. I wanted to hear *Boléro* again.

Helen Keller famously said that if she had to choose between being deaf and being blind, she'd be blind, because while blindness cut her off from things, deafness cut her off from people. For centuries the best available hearing aid was a horn, or ear trumpet, which people held to their ears to funnel in sound. In 1952 the first electronic hearing aid was developed. It worked by blasting amplified sound into a damaged ear. However, it (and the more advanced models that followed) could help only if the user had some residual hearing ability, just as glasses can help only those who still have some vision. Cochlear implants, on the other hand, bypass most of the ear's natural hearing mechanisms. The device's electrodes directly stimulate nerve endings in the ear, which transmit sound information to the brain. Since the surgery can eliminate any remaining hearing, implants are approved for use only in people who can't be helped by hearing aids. The first modern cochlear implants went on the market in Australia in 1982, and by 2004 approximately 82,500 people worldwide had been fitted with one.

When technicians activated my cochlear implant in October 2001, they gave me a pager-sized processor that decoded sound and sent it to a headpiece that clung magnetically to the implant underneath my skin. The headpiece contained a radio transmitter, which sent the processor's data to the implant at roughly 1 megabit per second. Sixteen electrodes curled up inside my cochlea strobed on and off to stimulate my auditory nerves. The processor's software gave me eight channels of auditory resolution, each representing a frequency range. The more channels the software delivers, the better the user can distinguish between sounds of different pitches.

Eight channels isn't much compared with the capacity of a normal ear, which has the equivalent of thirty-five hundred channels. Still, eight works well enough for speech, which doesn't have much pitch variation. Music is another story. The lowest of my eight channels captured everything from 250 hertz (about middle C on the piano) to 494 hertz (close to the B above middle C), making it nearly impossible for me to distinguish among the eleven notes in that range. Every note that fell into a particular channel sounded the same to me.

So in mid-2002, nine months after activation, I upgraded to a program called Hi-Res, which gave me sixteen channels — double the resolution! An audiologist plugged my processor into her laptop and uploaded the new code. I suddenly had a better ear, without surgery. In theory, I would now be able to distinguish among tones five notes apart instead of eleven.

I eagerly plugged my Walkman into my processor and turned it on. *Boléro* did sound better. But after a day or two, I realized that "better" still wasn't good enough. The improvement was small, like being in that art gallery again and seeing only a gleam of pink here, a bit of blue there. I wasn't hearing the *Boléro* I remembered.

At a cochlear implant conference in 2003, I heard Jay Rubinstein, a surgeon and researcher at the University of Washington, say that it took at least one hundred channels of auditory information to make music pleasurable. My jaw dropped. No wonder. I wasn't even close.

A year later I met Rubinstein at another conference, and he mentioned that there might be ways to bring music back to me. He told me about something called stochastic resonance; studies suggested that my music perception might be aided by deliberately adding noise to what I hear. He took a moment to give me a lesson in neural physiology. After a neuron fires, it goes dormant for a fraction of a second while it resets. During that phase, it misses any information that comes along. When an electrode zaps thousands of neurons at once, it forces them all to go dormant, making it impossible for them to receive pulses until they reset. That synchrony means I miss bits and pieces of information.

Desynchronizing the neurons, Rubinstein explained, would guarantee that they're never all dormant simultaneously. And the best way to get them out of sync is to beam random electrical noise at them. A few months later Rubinstein arranged a demonstration.

An audiologist at the University of Iowa working with Rubinstein handed me a processor loaded with the stochastic-resonance software. The first thing I heard was a loud whoosh — the random noise. It sounded like a cranked-up electric fan. But in about thirty seconds, the noise went away. I was puzzled. "You've adapted to it," the technician told me. The nervous system can habituate to any kind of everyday sound, but it adjusts especially quickly to noise with no variation. Stochastic-resonance noise is so content-free that the brain tunes it out in seconds.

In theory, the noise would add just enough energy to incoming sound to make faint details audible. In practice, everything I heard became rough and gritty. My own voice sounded vibrato, mechanical, and husky — even a little querulous, as if I were perpetually whining.

We tried some quick tests to take my newly programmed ear out for a spin. It performed slightly better in some ways, slightly worse in others — but there was no dramatic improvement. The audiologist wasn't surprised. She told me that in most cases a test subject's brain will take weeks or even months to make sense of the additional information. Furthermore, the settings she chose were only an educated guess at what might work for my particular physiology. Everyone is different. Finding the right setting is like fishing for one particular cod in the Atlantic.

The university loaned me the processor to test for a few months. As soon as I was back in the hotel, I tried my preferred version of *Boléro,* a 1982 recording conducted by Charles Dutoit with the Montreal Symphony Orchestra. It sounded different but not better.

Sitting at my keyboard, I sighed a little and tapped out an e-mail thanking Rubinstein and encouraging him to keep working on it.

Music depends on low frequencies for its richness and mellowness. The lowest-pitched string on a guitar vibrates at 83 hertz, but my Hi-Res software, like the eight-channel model, bottoms out at 250 hertz. I do hear something when I pluck a string, but it's not actually an 83-hertz sound. Even though the string is vibrating 83 times per second, portions of it are vibrating faster, giving rise to higher-frequency notes called harmonics. The harmonics are what I hear.

The engineers haven't gone below 250 hertz because the world's low-pitched sounds — air conditioners, engine rumbles — interfere with speech perception. Furthermore, increasing the total frequency range means decreasing resolution, because each channel has to accommodate more frequencies. Since speech perception has been the main goal during decades of research, the engineers didn't give much thought to representing low frequencies. Until Philip Loizou came along.

Loizou and his team of postdocs at the University of Texas at Dallas are trying to figure out ways to give cochlear-implant users access to more low frequencies. A week after my frustratingly inconclusive encounter with stochastic resonance, I traveled to Dallas and asked Loizou why the government would give him a grant to develop software that increases musical appreciation. "Music lifts up people's spirits, helps them forget things," he told me in his mild Greek accent. "The goal is to have the patient live a normal life, not to be deprived of anything."

Loizou is trying to negotiate a tradeoff: narrowing low-frequency channels while widening higher-frequency channels. But his theories only hinted at what specific configurations might work best, so Loizou was systematically trying a range of settings to see which ones got the better results.

The team's software ran only on a desktop computer, so on my visit to Dallas I had to be plugged directly into the machine. After a round of testing, a postdoc assured me, they would run *Boléro* through their software and pipe it into my processor via Windows Media Player.

I spent two and a half days hooked up to the computer, listening to endless sequences of tones — none of it music — in a windowless cubicle. Which of two tones sounded lower? Which of two versions of "Twinkle, Twinkle, Little Star" was more recognizable? Did this string of notes sound like a march or a waltz? It was exacting, high-concentration work — like taking an eye exam that lasted for two days. My responses produced reams of data that they would spend hours analyzing.

Forty minutes before my cab back to the airport was due, we finished the last test, and the postdoc fired up the programs he needed to play *Boléro.* Some of the lower pitches I'd heard in the previous two days had sounded rich and mellow, and I began thinking wistfully about those bassoons and oboes. I felt a rising sense of anticipation and hope.

I waited while the postdoc tinkered with the computer. And waited. Then I noticed the frustrated look of a man trying to get Windows to behave. "I do this all the time," he said, half to himself. Windows Media Player wouldn't play the file.

I suggested rebooting and sampling *Boléro* through a microphone. But the postdoc told me he couldn't do that in time for my plane. A later flight wasn't an option; I had to be back in the Bay Area. I was crushed. I walked out of the building with my shoulders slumped. Scientifically, the visit was a great success. But for me it was a failure. On the flight home, I plugged myself into my laptop and listened sadly to *Boléro* with Hi-Res. It was like eating cardboard.

It's June 2005, a few weeks after my visit to Dallas, and I'm ready to try again. A team of engineers at Advanced Bionics, one of three companies in the world that makes bionic ears, is working on a new software algorithm for so-called virtual channels. I hop on a flight to their Los Angeles headquarters, my CD player in hand.

My implant has sixteen electrodes, but the virtual-channels software will make my hardware act as if there are actually one hundred twenty-one. Manipulating the flow of electricity to target neurons between each electrode creates the illusion of seven new electrodes between each actual pair, similar to the way an audio engineer can make a sound appear to emanate from between two speakers. Jay Rubinstein had told me two years ago that it would take at least one hundred channels to create good music perception. I'm about to find out if he's right.

I'm sitting across a desk from Gulam Emadi, an Advanced Bionics researcher. He and an audiologist are about to fit me with the new software. Leo Litvak, who has spent three years developing the program, comes in to say hello. He's one of those people of whom others often say, "If Leo can't do it, it probably can't be done." And yet it would be hard to find a more modest person. Were it not for his clothes, which mark him as an Orthodox Jew, he would simply disappear in a roomful of people. Litvak tilts his head and smiles hello, shyly glances at Emadi's laptop, and sidles out.

At this point, I'm rationing my emotions like Spock. Hi-Res was a disappointment. Stochastic resonance remains a big if. The low-frequency experiment in Dallas was a bust. Emadi dinks with his computer and hands me my processor with the new software in it. I plug it into myself, plug my CD player into it, and press Play.

Boléro starts off softly and slowly, meandering like a breeze through the trees. *Da-da-da-dum, da-da-da-dum, dum-dum, da-da-da-dum.* I close my eyes to focus, switching between Hi-Res and the new software every twenty or thirty seconds by thumbing a blue dial on my processor.

My God, the oboe d'amores *do* sound richer and warmer. I let out a long, slow breath, coasting down a river of sound, waiting for the soprano saxophones and the piccolos. They come in around six minutes into the piece — and it's only then that I'll know if I've truly got it back.

As it turns out, I couldn't have chosen a better piece of music for testing new implant software. Some biographers have suggested that *Boléro*'s obsessive repetition is rooted in the neurological problems Ravel had started to exhibit in 1927, a year before he composed the piece. It's still up for debate whether he had early-onset Alzheimer's, a left-hemisphere brain lesion, or something else.

But *Boléro*'s obsessiveness, whatever its cause, is just right for my deafness. Over and over the theme repeats, allowing me to listen for specific details in each cycle.

At 5:59, the soprano saxophones leap out bright and clear, arcing above the snare drum. I hold my breath.

At 6:39, I hear the piccolos. For me, the stretch between 6:39 and 7:22 is the most *Boléro* of *Boléro,* the part I wait for each time. I concentrate. It sounds . . . *right.*

Hold on. Don't jump to conclusions. I backtrack to 5:59 and switch to Hi-Res. That heart-stopping leap has become an asthmatic whine. I backtrack again and switch to the new software. And there it is again, that exultant ascent. I can hear *Boléro*'s force, its intensity and passion. My chin starts to tremble.

I open my eyes, blinking back tears. "Congratulations," I say to Emadi. "You have done it." And I reach across the desk with absurd formality and shake his hand.

There's more technical work to do, more progress to be made, but I'm completely shattered. I keep zoning out and asking Emadi to repeat things. He passes me a box of tissues. I'm overtaken by a vast sensation of surprise. I did it. For years I pestered researchers and asked questions. Now I'm running 121 channels and I can hear music again.

That evening, in the airport, sitting numbly at the gate, I listen to *Boléro* again. I'd never made it through more than three or four minutes of the piece on Hi-Res before getting bored and turning it off. Now, I listen to the end, following the narrative, hearing again its holy madness.

I pull out the Advanced Bionics T-shirt that the team gave me and dab at my eyes.

During the next few days I walk around in a haze of disbelief, listening to *Boléro* over and over to prove to myself that I really am hearing it again. But *Boléro* is just one piece of music. Jonathan Berger, head of Stanford's music department, tells me in an e-mail, "There's not much of interest in terms of structure — it's a continuous crescendo, no surprises, no subtle interplay between development and contrast."

"In fact," he continues, "Ravel was not particularly happy that this study in orchestration became his big hit. It pales in comparison to any of his other music in terms of sophistication, innovation, grace, and depth."

So now it's time to try out music with sophistication, innovation, grace, and depth. But I don't know where to begin. I need an expert with first-rate equipment, a huge music collection, and the ability to pick just the right pieces for my newly reprogrammed ear. I put the question to craigslist: "Looking for a music geek." Within hours I hear from Tom Rettig, a San Francisco music producer.

In his studio, Rettig plays me Ravel's String Quartet in F Major and Philip Glass's String Quartet No. 5. I listen carefully, switching between the old software and the new. Both compositions sound enormously better on 121 channels. But when Rettig plays music with vocals, I discover that having 121 channels hasn't solved all my problems. While the crescendos in Dulce Pontes's *Canção do Mar* sound louder and clearer, I hear only white noise when her voice comes in. Rettig figures that relatively simple instrumentals are my best bet — pieces in which the instruments don't overlap too much — and that flutes and clarinets work well for me. Cavalcades of brass tend to overwhelm me and confuse my ear.

And some music just leaves me cold: I can't even get through Kraftwerk's *Tour de France.* I wave impatiently to Rettig to move on. (Later a friend tells me it's not the software — Kraftwerk is just dull. It makes me think that for the first time in my life I might be developing a taste in music.)

Listening to *Boléro* more carefully in Rettig's studio reveals other bugs. The drums sound squeaky — how can drums squeak? — and in the frenetic second half of the piece, I still have trouble separating the instruments.

After I get over the initial awe of hearing music again, I discover that it's harder for me to understand ordinary speech than it was before I went to virtual channels. I report this to Advanced Bionics, and my complaint is met by a rueful shaking of heads. I'm not the first person to say that, they tell me. The idea of virtual channels is a breakthrough, but the technology is still in the early stages of development.

But I no longer doubt that incredible things can be done with that unused 90 percent of my implant's hardware capacity. Tests conducted a month after my visit to Advanced Bionics show that my ability to discriminate among notes has improved considerably. With Hi-Res, I was able to identify notes only when they were at least 70 hertz apart. Now, I can hear notes that are only 30 hertz apart. It's like going from being able to tell the difference between red and blue to being able to distinguish between aquamarine and cobalt.

My hearing is no longer limited by the physical circumstances of my body. While my friends' ears will inevitably decline with age, mine will only get better.

DANIEL C. DENNETT

Show Me the Science

FROM *The New York Times*

PRESIDENT BUSH, announcing this month that he was in favor of teaching about "intelligent design" in the schools, said, "I think that part of education is to expose people to different schools of thought." A couple of weeks later, Senator Bill Frist of Tennessee, the Republican leader, made the same point. Teaching both intelligent design and evolution "doesn't force any particular theory on anyone," Frist said. "I think in a pluralistic society that is the fairest way to go about education and training people for the future."

Is "intelligent design" a legitimate school of scientific thought? Is there something to it, or have these people been taken in by one of the most ingenious hoaxes in the history of science? Wouldn't such a hoax be impossible? No. Here's how it has been done.

First, imagine how easy it would be for a determined band of naysayers to shake the world's confidence in quantum physics — how weird it is! — or Einsteinian relativity. In spite of a century of instruction and popularization by physicists, few people ever really get their heads around the concepts involved. Most people eventually cobble together a justification for accepting the assurances of the experts: "Well, they pretty much agree with one another, and they claim that it is their understanding of these strange topics that allows them to harness atomic energy, and to make transistors and lasers, which certainly do work . . ."

Fortunately for physicists, there is no powerful motivation for such a band of mischief-makers to form. They don't have to spend much time persuading people that quantum physics and Einsteinian relativity really have been established beyond all reasonable doubt.

With evolution, however, it is different. The fundamental scientific idea of evolution by natural selection is not just mind-boggling; natural selection, by executing God's traditional task of designing and creating all creatures great and small, also seems to deny one of the best reasons we have for believing in God. So there is plenty of motivation for resisting the assurances of the biologists. Nobody is immune to wishful thinking. It takes scientific discipline to protect ourselves from our own credulity, but we've also found ingenious ways to fool ourselves and others. Some of the methods used to exploit these urges are easy to analyze; others take a little more unpacking.

A creationist pamphlet sent to me some years ago had an amusing page in it, purporting to be part of a simple questionnaire:

> Test Two
> Do you know of any building that didn't have a builder? [YES] [NO]
> Do you know of any painting that didn't have a painter? [YES] [NO]
> Do you know of any car that didn't have a maker? [YES] [NO]
> If you answered YES for any of the above, give details.

Take that, you Darwinians! The presumed embarrassment of the test-taker when faced with this task perfectly expresses the incredulity many people feel when they confront Darwin's great idea. It seems obvious, doesn't it, that there couldn't be any designs without designers, any such creations without a creator.

Well, yes — until you look at what contemporary biology has demonstrated beyond all reasonable doubt: that natural selection — the process in which reproducing entities must compete for finite resources and thereby engage in a tournament of blind trial and error from which improvements automatically emerge — has the power to generate breathtakingly ingenious designs.

Take the development of the eye, which has been one of the favorite challenges of creationists. How on earth, they ask, could that engineering marvel be produced by a series of small, unplanned steps? Only an intelligent designer could have created such a brilliant arrangement of a shape-shifting lens, an aperture-adjusting iris, a light-sensitive image surface of exquisite sensitivity, all housed in a sphere that can shift its aim in a hundredth of a second and send megabytes of information to the visual cortex every second for years on end.

But as we learn more and more about the history of the genes involved, and how they work — all the way back to their predecessor genes in the sightless bacteria from which multicelled animals evolved more than a half-billion years ago — we can begin to tell the story of how photosensitive spots gradually turned into light-sensitive craters that could detect the direction from which light came and then gradually acquired their lenses, improving their information-gathering capacities all the while.

We can't yet say what all the details of this process were, but real eyes representative of all the intermediate stages can be found, dotted around the animal kingdom, and we have detailed computer models to demonstrate that the creative process works just as the theory says.

All it takes is a rare accident that gives one lucky animal a mutation that improves its vision over that of its siblings; if this helps it have more offspring than its rivals, this gives evolution an opportunity to raise the bar and ratchet up the design of the eye by one mindless step. And since these lucky improvements accumulate — this was Darwin's insight — eyes can automatically get better and better and better, without any intelligent designer.

Brilliant as the design of the eye is, it betrays its origin with a telltale flaw: the retina is inside out. The nerve fibers that carry the signals from the eye's rods and cones (which sense light and color) lie on top of them and have to plunge through a large hole in the retina to get to the brain, creating the blind spot. No intelligent designer would put such a clumsy arrangement in a camcorder, and this is just one of hundreds of accidents frozen in evolutionary history that confirm the mindlessness of the historical process.

If you still find Test Two compelling, a sort of cognitive illusion that you can feel even as you discount it, you are like just about everybody else in the world; the idea that natural selection has the power to generate such sophisticated designs is deeply counterintuitive. Francis Crick, one of the discoverers of DNA, once jokingly credited his colleague Leslie Orgel with "Orgel's Second Rule": evolution is cleverer than you are. Evolutionary biologists are often startled by the power of natural selection to "discover" an "ingenious" solution to a design problem posed in the lab.

This observation lets us address a slightly more sophisticated version of the cognitive illusion presented by Test Two. When evolutionists like Crick marvel at the cleverness of the process of natural selection, they are not acknowledging intelligent design. The designs found in nature are nothing short of brilliant, but the process of design that generates them is utterly lacking in intelligence of its own.

Intelligent-design advocates, however, exploit the ambiguity between process and product that is built into the word "design." For them the presence of a finished product (a fully evolved eye, for instance) is evidence of an intelligent design process. But this tempting conclusion is just what evolutionary biology has shown to be mistaken.

Yes, eyes are for seeing, but these and all the other purposes in the natural world can be generated by processes that are themselves without purposes and without intelligence. This is hard to understand, but so is the idea that colored objects in the world are composed of atoms that are not themselves colored and that heat is not made of tiny hot things.

The focus on intelligent design has, paradoxically, obscured something else: that genuine scientific controversies about evolution abound. In just about every field there are challenges to one established theory or another. The legitimate way to stir up such a storm is to come up with an alternative theory that makes a prediction that is crisply denied by the reigning theory — but that turns out to be true, or that explains something that has been baffling defenders of the status quo, or that unifies two distant theories at the cost of some element of the currently accepted view.

To date, the proponents of intelligent design have not produced anything like that. No experiments with results that challenge any mainstream biological understanding. No observations from the fossil record or genomics or biogeography or comparative anatomy that undermine standard evolutionary thinking.

Instead, the proponents of intelligent design use a ploy that works something like this. First you misuse or misdescribe some scientist's work. Then you get an angry rebuttal. Then, instead of dealing forthrightly with the charges leveled, you cite the rebuttal as evidence that there is a "controversy" to teach.

Note that the trick is content-free. You can use it on any topic. "Smith's work in geology supports my argument that the earth is flat," you say, misrepresenting Smith's work. When Smith responds with a denunciation of your misuse of her work, you respond, saying something like: "See what a controversy we have here? Professor Smith and I are locked in a titanic scientific debate. We should teach the controversy in the classrooms." And here is the delicious part: you can often exploit the very technicality of the issues to your own advantage, counting on most of us to miss the point in all the difficult details.

William Dembski, one of the most vocal supporters of intelligent design, notes that he provoked Thomas Schneider, a biologist, into a response that Dembski characterizes as "some hairsplitting that could only look ridiculous to outsider observers." What looks to scientists — and is — a knockout objection by Schneider is portrayed to most everyone else as ridiculous hairsplitting.

In short, no science. Indeed, no intelligent-design hypothesis has even been ventured as a rival explanation of any biological phenomenon. This might seem surprising to people who think that intelligent design competes directly with the hypothesis of nonintelligent design by natural selection. But saying, as intelligent-design proponents do, "You haven't explained everything yet" is not a competing hypothesis. Evolutionary biology certainly hasn't explained everything that perplexes biologists. But intelligent design hasn't yet tried to explain anything.

To formulate a competing hypothesis, you have to get down in the trenches and offer details that have testable implications. So far, intelligent-design proponents have conveniently sidestepped that requirement, claiming that they have no specifics in mind about who or what the intelligent designer might be.

To see this shortcoming in relief, consider an imaginary hypothesis of intelligent design that could explain the emergence of human beings on this planet:

> About six million years ago, intelligent genetic engineers from another galaxy visited Earth and decided that it would be a more interesting planet if there was a language-using, religion-forming species on it, so they sequestered some primates and genetically reengineered them to give them the language instinct and enlarged frontal lobes for planning and reflection. It worked.

If some version of this hypothesis were true, it could explain how and why human beings differ from their nearest relatives, and it would disconfirm the competing evolutionary hypotheses that are being pursued. We'd still have the problem of how these intelligent genetic engineers came to exist on their home planet, but we can safely ignore that complication for the time being, since there is not the slightest shred of evidence in favor of this hypothesis.

But here is something the intelligent-design community is reluctant to discuss: no other intelligent-design hypothesis has anything more going for it. In fact, my farfetched hypothesis has the advantage of being testable in principle: we could compare the human and chimpanzee genomes, looking for unmistakable signs of tampering by those genetic engineers from another galaxy. Finding some sort of user's manual neatly embedded in the apparently functionless "junk DNA" that makes up most of the human genome would be a Nobel Prize–winning coup for the intelligent-design gang, but if they are looking at all, they haven't come up with anything to report.

It's worth pointing out that there are plenty of substantive scientific controversies in biology that are not yet in the textbooks or the classrooms. The scientific participants in these arguments vie for acceptance among the relevant expert communities in peer-reviewed journals, and the writers and editors of textbooks grapple with judgments about which findings have risen to the level of acceptance — not yet truth — to make them worth serious consideration by undergraduates and high school students.

So get in line, intelligent designers. Get in line behind the hypothesis that life started on Mars and was blown here by a cosmic impact. Get in line behind the aquatic ape hypothesis, the gestural origin of language hypothesis, and the theory that singing came before language, to mention just a few of the enticing hypotheses that are actively defended but still insufficiently supported by hard facts.

The Discovery Institute, the conservative organization that has helped to put intelligent design on the map, complains that its members face hostility from the established scientific journals. But establishment hostility is not the real hurdle to intelligent design. If intelligent design were a scientific idea whose time had come, young scientists would be dashing around their labs, vying to win the Nobel Prizes that surely are in store for anybody who can overturn any significant proposition of contemporary evolutionary biology.

Remember cold fusion? The establishment was incredibly hostile to that hypothesis, but scientists around the world rushed to their labs in the effort to explore the idea, in hopes of sharing in the glory if it turned out to be true.

Instead of spending more than one million dollars a year on publishing books and articles for nonscientists and on other public relations efforts, the Discovery Institute should finance its own peer-reviewed electronic journal. That way the organization could live up to its self-professed image: the doughty defenders of brave iconoclasts bucking the establishment.

For now, though, the theory they are promoting is exactly what George Gilder, a long-time affiliate of the Discovery Institute, has said it is: "Intelligent design itself does not have any content."

Since there is no content, there is no "controversy" to teach about in biology class. But here is a good topic for a high school course on current events and politics: Is intelligent design a hoax? And if so, how was it perpetrated?

FRANS B. M. DE WAAL

How Animals Do Business

FROM *Scientific American*

JUST AS MY OFFICE would not stay empty for long were I to move out, nature's real estate changes hands all the time. Potential homes range from holes drilled by woodpeckers to empty shells on the beach. A typical example of what economists call a "vacancy chain" is the housing market among hermit crabs. To protect its soft abdomen, each crab carries its house around, usually an abandoned gastropod shell. The problem is that the crab grows, whereas its house does not. Hermit crabs are always on the lookout for new accommodations. The moment they upgrade to a roomier shell, other crabs line up for the vacated one.

One can easily see supply and demand at work here, but because it plays out on a rather impersonal level, few would view the crab version as related to human economic transactions. The crab interactions would be more interesting if the animals struck deals along the lines of "you can have my house if I can have that dead fish." Hermit crabs are not deal makers, though, and in fact have no qualms about evicting homeowners by force. Other, more social animals do negotiate, however, and their approach to the exchange of resources and services helps us understand how and why human economic behavior may have evolved.

The New Economics

Classical economics views people as profit maximizers driven by pure selfishness. As the seventeenth-century English philosopher Thomas Hobbes put it, "Every man is presumed to seek what is good for himselfe naturally, and what is just, only for Peaces sake, and accidentally." In this still prevailing view, sociality is but an afterthought, a "social contract" that our ancestors entered into because of its benefits, not because they were attracted to one another. For the biologist, this imaginary history falls as wide of the mark as can be. We descend from a long line of group-living primates, meaning that we are naturally equipped with a strong desire to fit in and find partners to live and work with. This evolutionary explanation for why we interact as we do is gaining influence with the advent of a new school, known as behavioral economics, that focuses on actual human behavior rather than on the abstract forces of the marketplace as a guide for understanding economic decision making. In 2002 the school was recognized by a shared Nobel Prize for two of its founders: Daniel Kahneman and Vernon L. Smith.

Animal behavioral economics is a fledgling field that lends support to the new theories by showing that basic human economic tendencies and preoccupations — such as reciprocity, the division of rewards, and cooperation — are not limited to our species. They probably evolved in other animals for the same reasons they evolved in us — to help individuals take optimal advantage of one another without undermining the shared interests that support group life.

Take a recent incident during my research at the Yerkes National Primate Research Center in Atlanta. We had taught capuchin monkeys to reach a cup of food on a tray by pulling on a bar attached to the tray. By making the tray too heavy for a single individual, we gave the monkeys a reason to work together.

On one occasion, the pulling was to be done by two females, Bias and Sammy. Sitting in adjoining cages, they successfully brought a tray bearing two cups of food within reach. Sammy, however, was in such a hurry to collect her reward that she released the bar and grabbed her cup before Bias had a chance to get hers. The tray bounced back, out of Bias's reach. While Sammy munched away, Bias threw a tantrum. She screamed her lungs out for half a minute until Sammy approached her pull bar again. She then helped Bias bring in the tray a second time. Sammy did not do so for her own benefit, because by now the cup accessible to her was empty.

Sammy's corrective behavior appeared to be a response to Bias's protest against the loss of an anticipated reward. Such action comes much closer to human economic transactions than that of the hermit crabs, because it shows cooperation, communication, and the fulfillment of an expectation, perhaps even a sense of obligation. Sammy seemed sensitive to the quid pro quo of the situation. This sensitivity is not surprising given that the group life of capuchin monkeys revolves around the same mixture of cooperation and competition that marks our own societies.

The Evolution of Reciprocity

Animals and people occasionally help one another without any obvious benefits for the helper. How could such behavior have evolved? If the aid is directed at a family member, the question is relatively easy to answer. "Blood is thicker than water," we say, and biologists recognize genetic advantages to such assistance: if your kin survive, the odds of your genes making their way into the next generation increase. But cooperation among unrelated individuals suggests no immediate genetic advantages. Pëtr Kropotkin, a Russian prince, offered an early explanation in his book *Mutual Aid,* published in 1902. If helping is communal, he reasoned, all parties stand to gain — everyone's chances for survival go up. We had to wait until 1971, however, for Robert L. Trivers, then at Harvard University, to phrase the issue in modern evolutionary terms with his theory of reciprocal altruism.

Trivers contended that making a sacrifice for another pays off if the other later returns the favor. Reciprocity boils down to "I'll scratch your back if you scratch mine." Do animals show such tit for tat? Monkeys and apes form coalitions; two or more individuals, for example, gang up on a third. And researchers have found a positive correlation between how often A supports B and how often B supports A. But does this mean that animals actually keep track of given and received favors? They may just divide the world into "buddies," whom they prefer, and "nonbuddies," whom they care little about. If such feelings are mutual, relationships will be either mutually helpful or mutually unhelpful. Such symmetries can account for the reciprocity reported for fish, vampire bats (which regurgitate blood to their buddies), dolphins, and many monkeys.

Just because these animals may not keep track of favors does not mean they lack reciprocity. The issue rather is how a favor done for another finds its way back to the original altruist. What exactly is the reciprocity mechanism? Mental record keeping is just one way of getting reciprocity to work, and whether animals do this remains to be tested. Thus far chimpanzees are the only exception. In the wild they hunt in teams to capture colobus monkeys. One hunter usually captures the prey, after which he tears it apart and shares it. Not everyone gets a piece, though, and even the highest-ranking male, if he did not take part in the hunt, may beg in vain. This by itself suggests reciprocity: hunters seem to enjoy priority during the division of spoils.

To try to find the mechanisms at work here, we exploited the tendency of these apes to share — which they also show in captivity — by handing one of the chimpanzees in our colony a watermelon or some branches with leaves. The owner would be at the center of a sharing cluster, soon to be followed by secondary clusters around individuals who had managed to get a major share, until all the food had trickled down to everyone. Claiming another's food by force is almost unheard of among chimpanzees — a phenomenon known as "respect of possession." Begging chimpanzees hold out their hand, palm upward, much like human beggars in the street. They whimper and whine, but aggressive confrontations are rare. If these do occur, it is almost always the possessor who initiates them to make someone leave the circle. She whacks the offenders over the head with a sizable branch or barks at them in a shrill voice until they leave her alone. Whatever their rank, possessors control the food flow.

We analyzed nearly seven thousand of these approaches, comparing the possessor's tolerance of specific beggars with previously received services. We had detailed records of grooming on the mornings of days with planned food tests. If the top male, Socko, had groomed May, for example, his chances of obtaining a few branches from her in the afternoon were much improved. This relation between past and present behavior proved general. Symmetrical connections could not explain this outcome, as the pattern varied from day to day. Ours was the first animal study to demonstrate a contingency between favors given and received. Moreover, these food-for-grooming deals were partner-specific — that is, May's tolerance benefited Socko, the one who had groomed her, but no one else.

This reciprocity mechanism requires memory of previous events as well as the coloring of memory such that it induces friendly behavior. In our own species, this coloring process is known as "gratitude," and there is no reason to call it something else in chimpanzees. Whether apes also feel obligations remains unclear, but it is interesting that the tendency to return favors is not the same for all relationships. Between individuals who associate and groom a great deal, a single grooming session carries little weight. All kinds of daily exchanges occur between them, probably without their keeping track. They seem instead to follow the buddy system discussed before. Only in the more distant relationships does grooming stand out as specifically deserving reward. Because Socko and May are not close friends, Socko's grooming was duly noticed.

A similar difference is apparent in human behavior, where we are more inclined to keep track of give-and-take with strangers and colleagues than with our friends and family. In fact, scorekeeping in close relationships, such as between spouses, is a sure sign of distrust.

Biological Markets

Because reciprocity requires partners, partner choice ranks as a central issue in behavioral economics. The hand-me-down housing of hermit crabs is exceedingly simple compared with the interactions among primates, which involve multiple partners exchanging multiple currencies, such as grooming, sex, support in fights, food, babysitting, and so on. This "marketplace of services," as I dubbed it in *Chimpanzee Politics,* means that each individual needs to be on good terms with higher-ups, to foster grooming partnerships, and — if ambitious — to strike deals with like-minded others. Chimpanzee males form coalitions to challenge the reigning ruler, a process fraught with risk. After an overthrow, the new ruler needs to keep his supporters contented: an alpha male who tries to monopolize the privileges of power, such as access to females, is unlikely to keep his position for long. And chimps do this without having read Niccolò Machiavelli.

With each individual shopping for the best partners and selling its own services, the framework for reciprocity becomes one of supply and demand, which is precisely what Ronald Noë and Peter Hammerstein, then at the Max Planck Institute for Behavioral

Physiology in Seewiesen, Germany, had in mind with their biological market theory. This theory, which applies whenever trading partners can choose with whom to deal, postulates that the value of commodities and partners varies with their availability. Two studies of market forces elaborate this point: one concerns the baby market among baboons, the other the job performance of small fish called cleaner wrasses.

Like all primate females, female baboons are irresistibly attracted to infants — not only their own but also those of others. They give friendly grunts and try to touch them. Mothers are highly protective, however, and reluctant to let anyone handle their precious newborns. To get close, interested females groom the mother while peeking over her shoulder or underneath her arm at the baby. After a relaxing grooming session, a mother may give in to the groomer's desire for a closer look. The other thus buys infant time. Market theory predicts that the value of babies should go up if there are fewer around. In a study of wild chacma baboons in South Africa, Louise Barrett of the University of Liverpool and Peter Henzi of the University of Central Lancashire, both in England, found that, indeed, mothers of rare infants were able to extract a higher price (longer grooming) than mothers in a troop full of babies.

Cleaner wrasses (*Labroides dimidiatus*) are small marine fish that feed on the external parasites of larger fish. Each cleaner owns a "station" on a reef, where clientele come to spread their pectoral fins and adopt postures that offer the cleaner a chance to do its job. The exchange exemplifies a perfect mutualism.

The cleaner nibbles the parasites off the client's body surface, gills, and even the inside of its mouth. Sometimes the cleaner is so busy that clients have to wait in line. Client fish come in two varieties: residents and roamers. Residents belong to species with small territories; they have no choice but to go to their local cleaner. Roamers, on the other hand, either hold large territories or travel widely, which means that they have several cleaning stations to choose from. They want short waiting times, excellent service, and no cheating. Cheating occurs when a cleaner fish takes a bite out of its client, feeding on healthy mucus. This makes clients jolt and swim away.

Research on cleaner wrasses by Redouan Bshary of the Max

Planck Institute in Seewiesen consists mainly of observations on the reef but also includes ingenious experiments in the laboratory. His papers read much like a manual for good business practice. Roamers are more likely to change stations if a cleaner has ignored them for too long or cheated them. Cleaners seem to know this and treat roamers better than they do residents. If a roamer and a resident arrive at the same time, the cleaner almost always services the roamer first. Residents have nowhere else to go, and so they can be kept waiting. The only category of fish that cleaners never cheat are predators, who possess a radical counterstrategy, which is to swallow the cleaner. With predators, cleaner fish wisely adopt, in Bshary's words, an "unconditionally cooperative strategy."

Biological market theory offers an elegant solution to the problem of freeloaders, which has occupied biologists for a long time because reciprocity systems are obviously vulnerable to those who take rather than give. Theorists often assume that offenders must be punished, although this has yet to be demonstrated for animals. Instead, cheaters can be taken care of in a much simpler way. If there is a choice of partners, animals can simply abandon unsatisfactory relationships and replace them with those offering more benefits. Market mechanisms are all that is needed to sideline profiteers. In our own societies, too, we neither like nor trust those who take more than they give, and we tend to stay away from them.

Fair Is Fair

To reap the benefits of cooperation, an individual must monitor its efforts relative to others and compare its rewards with the effort put in. To explore whether animals actually carry out such monitoring, we turned again to our capuchin monkeys, testing them in a miniature labor market inspired by field observations of capuchins attacking giant squirrels. Squirrel hunting is a group effort, but one in which all rewards end up in the hands of a single individual: the captor. If captors were to keep the prey solely for themselves, one can imagine that others would lose interest in joining them in the future. Capuchins share meat for the same reason chimpanzees (and people) do: there can be no joint hunting without joint payoffs.

We mimicked this situation in the laboratory by making certain that only one monkey (whom we called the winner) of a tray-pulling pair received a cup with apple pieces. Its partner (the laborer) had no food in its cup, which was obvious from the outset because the cups were transparent. Hence, the laborer pulled for the winner's benefit. The monkeys sat side by side, separated by mesh. From previous tests we knew that a food possessor might bring food to the partition and permit the neighbor to reach for it through the mesh. On rare occasions, the possessor pushes pieces to the other.

We contrasted collective pulls with solo pulls. In one condition, both animals had a pull bar and the tray was heavy; in the other, the partner lacked a bar and the winner handled a lighter tray on its own. We counted more acts of food sharing after collective than after solo pulls: winners were in effect compensating their partners for the assistance they had received. We also confirmed that sharing affects future cooperation. Because a pair's success rate would drop if the winner failed to share, paying the laborer was a smart strategy.

Sarah F. Brosnan, one of my colleagues at Yerkes, went further in exploring reactions to the way rewards are divided. She would offer a capuchin monkey a small pebble, then hold up a slice of cucumber as enticement for returning the pebble. The monkeys quickly grasped the principle of exchange. Placed side by side, two monkeys would gladly exchange pebbles for cucumber with the researcher. If one of them got grapes, however, and the other got cucumber, things took an unexpected turn. Grapes are much preferred. Monkeys who had been perfectly willing to work for cucumber suddenly went on strike. Not only did they perform reluctantly, seeing that the other was getting a better deal, but they became agitated, hurling the pebbles out of the test chamber and sometimes even the cucumber slices. A food normally never refused had become less than desirable.

To reject unequal pay — which people do as well — goes against the assumptions of traditional economics. If maximizing benefits were all that mattered, one should take what one can get and never let resentment or envy interfere. Behavioral economists, on the other hand, assume evolution has led to emotions that preserve the spirit of cooperation and that such emotions powerfully influence behavior. In the short run, caring about what others get may seem irrational, but in the long run it keeps one from being taken advantage of. Discouraging exploitation is critical for continued cooperation.

It is a lot of trouble, though, to always keep a watchful eye on the flow of benefits and favors. This is why humans protect themselves against freeloading and exploitation by forming buddy relationships with partners — such as spouses and good friends — who have withstood the test of time. Once we have determined whom to trust, we relax the rules. Only with more distant partners do we keep mental records and react strongly to imbalances, calling them "unfair."

We found indications for the same effect of social distance in chimpanzees. Straight tit for tat, as we have seen, is rare among friends who routinely do favors for one another. These relationships also seem relatively immune to inequity. Brosnan conducted her exchange task using grapes and cucumbers with chimpanzees as well as capuchins. The strongest reaction among chimpanzees concerned those who had known one another for a relatively short time, whereas the members of a colony that had lived together for more than thirty years hardly reacted at all. It is possible that the greater their familiarity, the longer the time frame over which chimpanzees evaluate their relationships. Only distant relations are sensitive to day-to-day fluctuations.

All economic agents, whether human or animal, need to come to grips with the freeloader problem and the way yields are divided after joint efforts. They do so by sharing most with those who help them most and by displaying strong emotional reactions to violated expectations. A truly evolutionary discipline of economics recognizes this shared psychology and considers the possibility that we embrace the golden rule not accidentally, as Hobbes thought, but as part of our background as cooperative primates.

DAVID DOBBS

Buried Answers

FROM *The New York Times Magazine*

WHEN DR. ALAN SCHILLER'S eighty-seven-year-old mother died in January, "it took some convincing," Schiller says, to get his siblings to agree to an autopsy. "They said, 'She had Alzheimer's. Let her rest.' But I told them, 'No, something seems funny to me. An autopsy is the only way to be sure.'" Schiller prevailed. A tanned, quick-minded, gregarious man in his sixties, he is naturally persuasive, and as chairman of pathology at Mount Sinai School of Medicine in New York he carries a certain authority regarding autopsies. The word "autopsy," he reminded his siblings, means to "see for oneself," and they should see what happened to their mother. Schiller's mother died in Miami, so he called his friend Dr. Robert Poppiti Jr., chairman of pathology at Mount Sinai Medical Center in Miami Beach. She was on the table the next morning.

In a living patient, Alzheimer's is a diagnosis of exclusion, one that ideally should be reached only by eliminating all testable causes of fading memory and mind. Confirming it requires directly examining the brain. The definitive markers are the tiny protein plaques and fibrous tangles that appear under the microscope in stained sections. But a good pathologist can spot advanced Alzheimer's just by looking at the whole brain. The brain will be shorter, front to back, and more squarish than normal — a reflection of Alzheimer's neuronal decimation, which shrinks the brain by up to 15 percent. That would presumably have been the case with Mrs. Schiller. But the autopsist found her brain to be of normal shape and size. Dissecting it, he discovered a half-dozen cystlike lesions scattered throughout — areas darker, softer, and less elastic than the buff-colored parts surrounding them.

"He called me right away," Schiller told me. "He said, 'Alan, your mother didn't have Alzheimer's. She had multi-infarct dementia.' You know what that is? It's a loss of mental capacity from a series of strokes. You can tell it because these cysts show up, little areas filled with fluid that comes in after a clot cuts off blood flow and the cells die. She had had multiple strokes. She didn't have Alzheimer's at all. She'd been slowly killed by strokes.

"Now, this is useful information," Schiller continued. "For one, it means I should worry less about getting Alzheimer's but maybe more about my cardiovascular health. It also means they could have been treating her for stroke. She might have had a very different life. But Alzheimer's has become a wastebasket diagnosis. You behave strangely and you're old, you have Alzheimer's. But other things, like this multi-infarct dementia, can produce the same symptoms. And no one ever checked for that."

This is the point that Schiller, a champion of the autopsy, means to make: even in today's high-tech medical world, the low-tech hospital autopsy — not the crime-oriented forensic autopsy glorified in television, but the routine autopsy done on patients who die in hospitals — provides a uniquely effective means of quality control and knowledge. It exposes mistakes and bad habits, evaluates diagnostic and treatment routines, and detects new disease. It is, Schiller says, the most powerful tool in the history of medicine, responsible for most of our knowledge of anatomy and disease, and it remains vital. "Neglecting the autopsy," he says, "is anathema to the whole practice of medicine."

Yet the hospital autopsy is neglected. When Schiller went to medical school in the 1960s, hospitals in the United States autopsied almost half of all deaths, and the autopsy was familiar to medical students and practitioners alike. The United States now does postmortems on fewer than 5 percent of hospital deaths, and the procedure is alien to almost every doctor trained in the last thirty years. Schiller has fought this. Soon after he took over Mount Sinai's pathology department sixteen years ago, a time when many hospitals were closing their autopsy facilities, he built what he calls "a beautiful new morgue," spending more than a million dollars. "I wanted a grand opening, a public thing," he says, laughing. "You know — ribbons, speeches. The hospital said, 'Are you nuts? It's a morgue!'" The hospital has backed him otherwise. By pushing clinicians to ask for autopsies and by doing good autopsies that quickly give clinicians useful feedback, Schiller has lifted Mount Sinai's autopsy rates from the single digits to the midteens. But only a few hospitals, almost all of them teaching hospitals, like Mount Sinai, still do that many. Elsewhere the autopsy is dying.

Dr. George Lundberg, a pathologist who edited the *Journal of the American Medical Association* from 1982 until 1999 and now edits the online medical journal *Medscape General Medicine,* has, like Schiller, spent much of his career trying to revive the autopsy. The heart of his plaint is that nothing reveals error as well as the autopsy. As Lundberg noted in a 1998 article, numerous studies over the last century have found that in 25 to 40 percent of cases in which an autopsy is done, it reveals an undiagnosed cause of death. Because of those errors, in 7 to 12 percent of the cases, treatment that might have been lifesaving wasn't prescribed. (In the other cases, the disease might have advanced beyond treatment or there might have been multiple causes of death.) These figures roughly match those found in the first discrepancy studies, done in the early 1910s. "No improvement!" Lundberg notes. "Low-tech autopsy trumps high-tech medicine . . . again and again."

Lundberg doesn't fantasize that the autopsy can make medicine mistake-free; medicine poses puzzles too various and complex to allow us to expect perfection, and indeed error rates are about the same no matter how many autopsies are done. But autopsies can keep doctors from *repeating* mistakes, and thus advance medicine. Doctors miss things. But without autopsies, they don't know when they've missed something fatal and so are likely to miss it again. They miss the chance to learn from their mistakes. Instead, they bury them. This, Lundberg says, "is endlessly galling."

As Lundberg sees it, "If you want to base your medicine on evidence, if you want to reduce error, if you simply want to know what you are doing, then you should start by evaluating the care given to your sickest patients — the ones who die."

The autopsy's intellectual founder was Giovanni Morgagni, a physician and professor at the University of Padua who wrote one of the most gruesome, humane, and riveting early texts of modern medicine, *The Seats and Causes of Disease Investigated by Anatomy.* Published in 1761, when Morgagni was seventy-nine, the book describes nearly seven hundred autopsies he performed. His lucid, compassionate accounts demonstrated irrefutably that illness works in traceable, physical ways; medicine, therefore, should be an empirical endeavor aimed at particular physical processes rather than "humors," spirits, or other intangibles.

Morgagni's perspective was carried into the present era by William Osler, a Canadian who practiced and taught medicine in the United States in the late 1800s. Osler exerted more influence on twentieth-century medicine than any other doctor, primarily by creating at Johns Hopkins Medical School the model for medical education still used today, with students seeing patients beginning in their third year and training in internships and residencies after graduating. Osler placed the autopsy at the center of this education, performing more than a thousand postmortems himself and insisting that staff members and students do them regularly. Tracking the necrotic footprints of their own missteps, he believed, would teach them lessons far more memorable than any text could.

Osler's argument was strengthened in the early 1910s by the work of Richard Clarke Cabot, who reviewed the records and autopsies of thousands of patients at Massachusetts General Hospital and found that the autopsies showed clinical diagnoses to be wrong about 40 percent of the time — a finding replicated many times since. His reports helped solidify the autopsy's central role in medical education and practice. Autopsy rates began to rise. By World War II, they were nearing 50 percent, and autopsies had become standard in medical schools and many hospitals, where weekly mortality and morbidity conferences often focused on what autopsies had revealed about the diagnosis and treatment of patients' illnesses.

That midcentury peak helped drive remarkable medical progress. In 1945, for instance, the chance of survival for a patient with an aortic aneurysm was little better than it was a century earlier. But in the fifties and sixties, people like the pioneering cardiovascular surgeon Michael DeBakey learned through trial and error — the errors offering their lessons only through autopsy — how to repair and replace first lower sections of the aorta in the abdomen and then, working up toward the heart, the biggest, most pressurized, and most vital sections. By 1960 aortic repairs were routine. By 1970 the lessons learned helped make open-heart surgery common as well.

Autopsies similarly advanced other areas of medicine. They played central roles in diagnosing and spurring treatment for sudden infant death syndrome, Legionnaires' disease, toxic-shock syndrome, hantavirus, HIV, Ebola, and other infectious diseases and helped make the association between lung cancer and smoking. These medical advances would have come about much more slowly without autopsies. In 1999, for instance, when four New York City residents died of what was diagnosed as St. Louis encephalitis, it was only because the city's unusually aggressive medical examiner's office insisted on autopsying them that they were discovered to be the first American victims of West Nile virus. (In most cities, the four would have been buried or cremated without autopsy.) The federal Centers for Disease Control and Prevention subsequently established a nationwide monitoring, control, and treatment system credited with preventing scores or perhaps hundreds of deaths.

Though yet deadlier pathogens, like those that cause avian flu, mad cow disease, and SARS, will almost certainly make their way to the United States, our low autopsy rates may well delay their detection. Prion diseases — for instance, mad cow, which in humans is called variant Creutzfeldt-Jakob disease — cause a neurological death that doctors could mistake for multi-infarct dementia, encephalitis, or even a fast-moving Alzheimer's. A patient who died of a prion disease might go unautopsied and be cremated or buried, leaving the prion disease and its source undetected. With some two hundred thousand Alzheimer's and stroke patients buried unautopsied each year, this may have already happened.

After twenty years of making arguments for autopsy, Lundberg says that he feels like the football coach in the joke about the dim and unmotivated player. "What's wrong with you, Jones?" the coach says. "Are you ignorant? Or just apathetic?" To which Jones replies, "I don't know, and I don't care."

When Lundberg talks autopsy to doctors' groups or health-care policymakers, his audiences generally agree that we should do more autopsies. "But nobody takes the steps to make it happen," Lundberg laments. They shake their heads in dismay, then return to business as usual. The forces arrayed against the autopsy — regulatory, economic, and cultural — seem to overcome any impulse to revive it.

It starts with pathologists. Most pathologists don't like autopsies. The procedure entails two to four smelly hours at the table and as many again analyzing samples, and the work comes atop other duties — ones that feel more urgent — like analyzing biopsies of living patients. Autopsies seldom advance careers or status, and most hospitals don't pay pathologists for doing them or provide updated equipment to ease the job and get the most out of the sampled tissues.

Hospitals say the problem is money. An autopsy can cost from $2,000 to $4,000, and insurance won't cover it. Most patient families blanch if asked to pay for an autopsy, and many can't afford to after paying medical and funeral bills. So the hospital gets the tab. For most of the postwar period up to 1970, hospitals generally paid it, essentially because they had to: the Joint Commission on Accreditation of Healthcare Organizations required hospitals to maintain autopsy rates of at least 20 percent (25 percent for teaching hospitals), which, then and now, is the rate most advocates say is the minimum for monitoring diagnostic and hospital error. The commission eliminated that requirement in 1970. Lundberg says that this happened because hospitals, which had already allowed the rate to drop to close to 20 percent since its 1950s high of about 50 percent, wanted to let it drop further and pressured the commission. The commission's current president, Dr. Dennis S. O'Leary, says it eliminated the standard because too many hospitals were doing poor autopsies — and often only the cheapest, simplest ones — just to make the quota. In any event, few hospitals have paid for autopsies since then. Money is too scarce, they say, and the needs of living patients too great.

But this argument fails scrutiny. For starters, hospitals do get money for autopsies: Medicare includes an autopsy allowance in the lump sum it pays hospitals for each Medicare inpatient, and those patients account for three-quarters of all hospital deaths. This money could easily finance double-digit autopsy rates. But most hospitals spend it on other things. Lundberg and others have urged the Department of Health and Human Services to make Medicare payments contingent on hospitals' meeting a certain autopsy rate. But the agency shows no interest in doing so.

The hospitals' dodge on this issue reveals less about finance than about attitude. They have the money. They don't use it for autopsies because they don't value autopsies. The hospitals that do — teaching hospitals like New York's Mount Sinai; Dartmouth-Hitchcock Medical Center, in Lebanon, New Hampshire; and Baylor University Medical Center, in Dallas — manage to absorb the costs. Their lobbies may not be as nice. But they have a much better idea where their errors are. "People sometimes ask me how good a hospital is," Lundberg says. "With most hospitals, the answer is that no one knows — because the hospital has no way to know how many and what kinds of mistakes they make."

Another oft-cited inhibitor is doctors' fears of being sued if an autopsy finds error. Research shows no link between autopsies and increased tendencies to sue. Patients' families say this is because doctors increase trust by asking for an autopsy and encourage suspicion and acrimony when they don't request one.

Doctors seem to overestimate families' resistance to autopsies. In one survey of doctors and families of seriously ill patients, 89 percent of the physicians said they planned to request autopsies and two-thirds of the families said they would probably grant permission. "But only 23 percent of them actually got autopsied," says Dr. Elizabeth Burton, a pathologist and autopsy advocate at Baylor who was one of the study's authors. "Why? We went back and asked the families. Many were never asked. Among those who were, the biggest deciding factor was how strongly the doctor recommended it. If the doctor showed conviction and a good reason, the families almost always went for it."

Some families do object, of course, and variations on the refrain of Schiller's siblings, "Let her rest," have answered many an autopsy request. But if the doctor persists and wins approval, the family often gains a welcome sense of resolution. An extreme example of how a postmortem can resolve troubling questions surfaced in the case of Terri Schiavo; the autopsy revealed far more precisely the extent of her brain damage, resolving that she was truly vegetative, as most doctors believed, rather than minimally conscious, aware, and responsive, as her parents believed. More commonly, results clarify family health issues. People who go through a miscarriage or parents whose children have died seem to benefit especially. Tracing death to a particular cause seems to ease anguish about things done or not done. Yet few doctors regularly ask to perform the postmortem.

Perhaps the most troubling reason for the decline of the autopsy is the overconfidence that doctors — and patients — have in MRIs and other high-tech diagnostic technologies. Bill Pellan of the Pinellas County medical examiner's office says, "We get this all the time. The doctor will get our report and call and say, 'But there can't be a lacerated aorta. We did a whole set of scans.' We have to remind him we held the heart in our hands." In fact, advanced diagnostic tools do miss critical problems and actually produce more false-negative diagnoses than older methods, probably because doctors accept results too readily. One study of diagnostic errors made from 1959 to 1989 (the period that brought us CAT scans, MRIs, and many other high-tech diagnostics) found that while false-positive diagnoses remained about 10 percent during that time, false-negative diagnoses — that is, the erroneous ruling out of a condition — rose from 24 percent to 34 percent. Another study found that errors occur at the same rate regardless of whether sophisticated diagnostic tools are used. Yet doctors routinely dismiss possible diagnoses because high-tech tools show negative results. One of my own family doctors told me that he rarely asks for an autopsy because "with MRIs and CAT scans and everything else, we usually know why they died."

This sense of omniscience, Lundberg says, is part of "a vast cultural delusion." At his most incensed, Lundberg says he feels that his fellow doctors simply don't want to face their own fallibility. But Lundberg's indictment is even broader. The autopsy's decline reflects not just individual arrogance, but also the general state of health care: the increasing distance and unease between doctors and patients and their families, a pervasive fear of lawsuits, our denial of age and death, and, especially, our credulous infatuation with technology. Our doctors' overconfidence, less bigheaded than blithe, is part of the medicine we've come to expect.

Recently I stood in the autopsy room of a large teaching hospital waiting for a body to be brought up from the morgue. The young pathologist who would be overseeing the autopsy told me what little he knew of the morning's patient. The middle-aged man had come to an emergency room suffering seizures. A CAT scan of his head showed a lesion, possibly a tumor, in his left frontal lobe. He initially refused a biopsy, saying that he might seek a second opinion. The emergency-room doctor, worried about pressure in the patient's skull if the mass expanded, put him on anti-inflammatory steroids and sent him home. Sometime later the man came in again with stronger and more persistent seizures. Despite efforts to ease pressure in his skull, he progressed from seizures to a coma and died. Midmorning the day after that he was on a gurney on his way to the autopsy room. The man was not overweight and had no known history of serious illness. His main compromising factors were that he was an ex–drug user and a smoker. "The drug use would suggest infection," the pathologist said. "The smoking, obviously, cancer."

So what killed him?

"Most likely he herniated," the pathologist said. "Things got too tight in his skull from whatever this growth was, and the pressure builds and finally it pushes the base of the brain down through the opening where the spinal cord enters the skull. That fits with the way he died. But even if that's right, we still don't know what the lesion is." At this point we heard the rumbling of wheels, and the autopsy assistant pushed a gurney covered with a canvas tent into the room. "We'll know more soon," the pathologist said.

He stepped out to get gowned up, and I went in to watch the assistant prepare things. By then the canvas tent had been removed to reveal a body wrapped in sheets. The assistant worked efficiently but with a calm, understated respect. With no more force than necessary, he pulled the body from the gurney onto the autopsy table and unwrapped it. The patient appeared to be thinking: his eyes, slightly open, stared dreamily at the ceiling.

In addition to the pathologist, the assistant, and a pathology resident, who would do the actual knife work, eight others attended, including a fourth-year medical student, two residents, three neuropathologists, and a cardiac pathologist who had just dissected another patient's heart and lingered to see how the brain case played out. As people milled and talked, the assistant sank a scalpel into the flesh behind the man's ear and began cutting a high arc behind the rear crown of the skull. When he reached the other ear, he pulled the scalp's flesh away from the skull a bit, crimped a towel over the front edge of the opening he had made and, using it for grip, pulled the scalp forward over the man's head. When he was done, the man's skull lay completely exposed and his inside-out scalp covered his face down to his mouth. Now a neuropathologist, wielding the skull saw (like a cordless kitchen mixer with a rotary blade), carefully cut a big oval in the rear and top of the man's skull. He then used a hammer and chisel to tap around the seam. Finally he tapped the chisel in at the top of the cut and pried. With a sucking sound the skull cap pulled away.

The brain looked unexpectedly smooth. "That's the swelling," the neuropathologist said. "The convolutions usually show much more plainly." He gently pulled back the frontal lobe and slipped scissors behind the eyes to snip the optical nerves, then the carotid arteries, and finally the spinal cord itself. Then he gently removed the brain and set it upside down on a table.

Even my untrained eyes could tell things weren't quite right: the left hemisphere was swollen. The growth in the left frontal lobe, less a lump than a slightly raised oval area about an inch long, was paler, yellower, firmer, and more granular than the pinkish-tan tissue surrounding it. "Could be a tumor," the neuropathologist said. "Could be an infection. We'll know more in a few days." Similar lesions were eventually identified on both sides of the brain.

With a pair of scissors, he pointed at a bulbous area around the brain stem. "Here's the herniation. See how it protrudes? That's where it got pushed down through the opening where the spinal cord comes through. That's the medulla that pushed through, which, among other things, controls the heart and breathing. That's just not consistent with life."

He clipped a few samples from the lesion, and with that he was largely done. The assistant, meanwhile, worked on, and with the brain exam finished, the pathologist soon joined him. They extracted meaty lungs and a big liver. Pus oozed out when the trachea was cut. All this suggested systemic infection. "At this point, I'd call it an even toss between infection and tumor," the pathologist said. "If he tests positive for HIV, my money goes on infection."

This initial exam of the organs took some fifteen minutes. When it was finished, the group spent another hour dissecting the organs. The exercise was now more educational than diagnostic, but the pathologist showed no sign of routine-induced boredom; on the contrary, he clearly enjoyed showing the residents the hidden adrenal glands, the chest-wall vessels sometimes used for coronary bypasses, and the vagus nerve's lacy, laddered course through the chest.

The full results would take several more days to come in. But they knew by the next day that the patient was HIV-positive, and by the second day that the mass was not cancerous but an infection found mainly in immunocompromised patients like this one.

These findings had multilayered implications. That the man had HIV, for instance, would presumably mean something to any of his sexual partners. (Many states require the primary physician to contact sexual partners in such cases.) The rest of his family might find some relief in knowing that there was no tumor and that their own cancer risk was thus not raised. Beyond that, the case's main epidemiological significance was its addition to evidence that infections form an ever-growing but oft-overlooked cause of death — another small correction in our assessment of what kills us. And that makes for better doctors. "You don't learn these things all at once," the pathologist said. "You learn a lot all at once in med school, sure. But after that, you become a better doctor by learning a little bit at a time. Incremental adjustments. That's what makes us better doctors. And this is the place you learn them better than anywhere else."

When a believer is in the full flush of describing autopsy's gifts, when you witness how quickly and effectively the procedure delivers them, it's easy to think that the autopsy will make a comeback. How could it not? At a time when medicine takes continuous fire regarding errors — politicians and patient advocates outraged at studies showing that one hundred thousand Americans die each year from medical errors, tort lawyers chasing mistakes on which to hang huge judgments, malpractice rates jumping at triple-digit rates — how can medicine ignore an instrument proven to detect error?

Yet it does. Other than hoping for a long shot, like Medicare or the Joint Commission on Accreditation of Healthcare Organizations setting autopsy-rate requirements, there is seemingly no quick return to routine autopsies. "We just have to do this one hospital at a time," says Dr. Pat Lento, head of the autopsy service at Mount Sinai in New York. But most hospitals have no plans to revive the autopsy. And while physician organizations like the AMA generally support the autopsy, most doctors don't avail themselves of it. The sad truth is that most of medicine seems to have relegated the postmortem to a cabinet of archaic tools, as if the body's direct lessons no longer mattered. In the end, the autopsy's troubles resemble those in a medical case in which the causes stand clear and a cure stands ready, but the patient doesn't take things seriously enough to pursue the fix.

Toward the end of the autopsy of the man who died from an ignored infection, someone asked the assistant if he could really put him back together for a funeral. It was almost two P.M. and the man was in pieces. His torso was a big red bowl formed by his back ribs, his skin hung splayed on either side, and his scalp was stretched inside-out over his face. The assistant smiled and said, "Oh, sure." The pathologist added, "Absolutely! This guy could go to his wake tonight."

And so it was. Unlike most things, an autopsied body can be put back together far more easily than it can be taken apart. It took less than half an hour to replace the breastplate and sew up the man's torso; if he had a suit, it would fit as before and hide all. The skull cap all but snapped into place. The assistant rolled the man's scalp back over his head and started to suture it up. When he was done, our patient looked pretty good indeed. It was remarkable, actually, after all we had found about what ailed him, that he should still gaze at the ceiling, unchanged and none the wiser.

MARK DOWIE

Conservation Refugees

FROM *Orion*

A LOW FOG ENVELOPS the steep and remote valleys of southwestern Uganda most mornings, as birds found only in this small corner of the continent rise in chorus and the great apes drink from clear streams. Days in the dense montane forest are quiet and steamy. Nights are an exaltation of insects and primate howling. For thousands of years the Batwa people thrived in this soundscape, in such close harmony with the forest that early-twentieth-century wildlife biologists who studied the flora and fauna of the region barely noticed their existence. They were, as one naturalist noted, "part of the fauna."

In the 1930s Ugandan leaders were persuaded by international conservationists that this area was threatened by loggers, miners, and other extractive interests. In response, three forest reserves were created — the Mgahinga, the Echuya, and the Bwindi — all of which overlapped with the Batwa's ancestral territory. For sixty years these reserves simply existed on paper, which kept them off-limits to extractors. And the Batwa stayed on, living as they had for generations, in reciprocity with the diverse biota that first drew conservationists to the region.

However, when the reserves were formally designated as national parks in 1991 and a bureaucracy was created and funded by the World Bank's Global Environment Facility to manage them, a rumor was in circulation that the Batwa were hunting and eating silverback gorillas, which by that time were widely recognized as a threatened species and also, increasingly, as a featured attraction for ecotourists from Europe and America. Gorillas were being disturbed and even poached, the Batwa admitted, but by Bahutu, Batutsi, Bantu, and other tribes who invaded the forest from outside villages. The Batwa, who felt a strong kinship with the great apes, adamantly denied killing them. Nonetheless, under pressure from traditional Western conservationists, who had come to believe that wilderness and human community were incompatible, the Batwa were forcibly expelled from their homeland.

These forests are so dense that the Batwa lost perspective when they first came out. Some even stepped in front of moving vehicles. Now they are living in shabby squatter camps on the perimeter of the parks, without running water or sanitation.

Tomas Mtwandi, who was born in the Mgahinga and evicted with his family when he was fourteen, is adapting slowly and reluctantly to modern life. He is employed as an indentured laborer for a local Bantu farmer and is raising a family in a one-room shack near the Bwindi park border. He is regarded as rich by his neighbors because his roof doesn't leak and he has a makeshift metal door on his mud-wall home. As a "registered resource user," Mtwandi is permitted to harvest honey from the Bwindi and pay an occasional visit to the graves of his ancestors in the Mgahinga, but he does so at the risk of being mistaken for a poacher and shot on sight by paid wardens from neighboring tribes. His forest knowledge is waning, and his family's nutrition is poor. In the forest they had meat, roots, fruit, and a balanced diet. Today they have a little money but no meat. In one more generation their forest-based culture — songs, rituals, traditions, and stories — will be gone.

It's no secret that millions of native peoples around the world have been pushed off their land to make room for big oil, big metal, big timber, and big agriculture. But few people realize that the same thing has happened for a much nobler cause: land and wildlife conservation. Today the list of culture-wrecking institutions put forth by tribal leaders on almost every continent includes not only Shell, Texaco, Freeport, and Bechtel, but also more surprising names like Conservation International (CI), The Nature Conservancy (TNC), the World Wildlife Fund (WWF), and the Wildlife Conservation Society (WCS). Even the more culturally sensitive World Conservation Union (IUCN) might get a mention.

In early 2004 a United Nations meeting was convened in New York for the ninth year in a row to push for passage of a resolution protecting the territorial and human rights of indigenous peoples. The UN draft declaration states: "Indigenous peoples shall not be forcibly removed from their lands or territories. No relocation shall take place without the free and informed consent of the indigenous peoples concerned and after agreement on just and fair compensation and, where possible, with the option to return." During the meeting an indigenous delegate who did not identify herself rose to state that while extractive industries were still a serious threat to their welfare and cultural integrity, their new and biggest enemy was "conservation."

Later that spring, at a meeting of the International Forum on Indigenous Mapping in Vancouver, British Columbia, all two hundred delegates signed a declaration stating that the "activities of conservation organizations now represent the single biggest threat to the integrity of indigenous lands." These are rhetorical jabs, of course, but they have shaken the international conservation community, as have a subsequent spate of critical articles and studies, two of them conducted by the Ford Foundation, calling big conservation to task for its historical mistreatment of indigenous peoples.

"We are enemies of conservation," declared Maasai leader Martin Saning'o, standing before a session of the November 2004 World Conservation Congress sponsored by IUCN in Bangkok, Thailand. The nomadic Maasai, who have over the past thirty years lost most of their grazing range to conservation projects throughout eastern Africa, hadn't always felt that way. In fact, Saning'o reminded his audience, "we were the original conservationists." The room was hushed as he quietly explained how pastoral and nomadic cattlemen have traditionally protected their range: "Our ways of farming pollinated diverse seed species and maintained corridors between ecosystems." Then he tried to fathom the strange version of land conservation that has impoverished his people, more than one hundred thousand of whom have been displaced from southern Kenya and the Serengeti Plains of Tanzania. Like the Batwa, the Maasai have not been fairly compensated. Their culture is dissolving and they live in poverty.

"We don't want to be like you," Saning'o told a room of shocked white faces. "We want you to be like us. We are here to change your minds. You cannot accomplish conservation without us."

Although he might not have realized it, Saning'o was speaking for a growing worldwide movement of indigenous peoples who think of themselves as conservation refugees. Not to be confused with ecological refugees — people forced to abandon their homelands as a result of unbearable heat, drought, desertification, flooding, disease, or other consequences of climate chaos — conservation refugees are removed from their lands involuntarily, either forcibly or through a variety of less coercive measures. The gentler, more benign methods are sometimes called "soft eviction" or "voluntary resettlement," though the latter is contestable. Soft or hard, the main complaint heard in the makeshift villages bordering parks and at meetings like the World Conservation Congress in Bangkok is that relocation often occurs with the tacit approval or benign neglect of one of the five big international nongovernmental conservation organizations, or, as they have been nicknamed by indigenous leaders, the BINGOs.

The rationale for "internal displacements," as these evictions are officially called, usually involves a perceived threat to the biological diversity of a large geographical area, variously designated by one or more of the BINGOs as an "ecological hot spot," an "ecoregion," a "vulnerable ecosystem," a "biological corridor," or a "living landscape." The huge parks and reserves that are created often involve a debt-for-nature swap (some of the host country's national debt paid off or retired in exchange for the protection of a parcel of sensitive land) or similar financial incentive provided by the World Bank's Global Environment Facility and one or more of its "executing agencies" (bilateral and multilateral banks). This trade may be paired with an offer made by the funding organization to pay for the management of the park or reserve. Broad rules for human use and habitation of the protected area are set and enforced by the host nation, often following the advice and counsel of a BINGO, which might even be given management powers over the area. Indigenous peoples are often left out of the process entirely.

Curious about this brand of conservation that puts the rights of nature before the rights of people, I set out last autumn to meet the issue face to face. I visited with tribal members on three continents who were grappling with the consequences of Western conservation and found an alarming similarity among the stories I heard.

Khon Noi, matriarch of a remote mountain village, huddles next to an open-pit stove in the loose, brightly colored clothes that identify her as a Karen, the most populous of six tribes found in the lush, mountainous reaches of far northern Thailand. Her village of sixty-five families has been in the same wide valley for over two hundred years. She chews betel, spitting its bright red juice into the fire, and speaks softly through black teeth. She tells me I can use her name, as long as I don't identify her village.

"The government has no idea who I am," she says. "The only person in the village they know by name is the 'headman' they appointed to represent us in government negotiations. They were here last week, in military uniforms, to tell us we could no longer practice rotational agriculture in this valley. If they knew that someone here was saying bad things about them they would come back again and move us out."

In a recent outburst of environmental enthusiasm stimulated by generous financial offerings from the Global Environment Facility, the Thai government has been creating national parks as fast as the Royal Forest Department can map them. Ten years ago there was barely a park to be found in Thailand, and because those few that existed were unmarked "paper parks," few Thais even knew they were there. Now there are 114 land parks and 24 marine parks on the map. Almost twenty-five thousand square kilometers, most of which are occupied by hill and fishing tribes, are now managed by the forest department as protected areas.

"Men in uniform just appeared one day, out of nowhere, showing their guns," Kohn Noi recalls, "and telling us that we were now living in a national park. That was the first we knew of it. Our own guns were confiscated . . . no more hunting, no more trapping, no more snaring, and no more 'slash and burn.' That's what they call our agriculture. We call it crop rotation and we've been doing it in this valley for over two hundred years. Soon we will be forced to sell rice to pay for greens and legumes we are no longer allowed to grow here. Hunting we can live without, as we raise chickens, pigs, and buffalo. But rotational farming is our way of life."

A week before our conversation, a short flight south of Noi's village, six thousand conservationists were attending the World Conservation Congress in Bangkok. Lining the hallways of a massive convention center were the display booths of big conservation, adorned with larger-than-life photos of indigenous peoples in splendid tribal attire. At huge plenary sessions praise was heaped on

Thailand's beloved Queen Sirikit and her environment minister, who came accompanied by a sizable delegation from the Royal Forest Department.

But if delegates had taken the time to attend smaller panels and workshops, some held outside the convention center in a parking lot, they would have heard Khon Noi's story repeated a dozen times or more by indigenous leaders who came to Bangkok from every continent, at great expense, to lobby conservation biologists and government bureaucrats for fairer treatment. And they would have heard a young Karen father of two boys ask why his country, whose cabinet had ordered its environmental bureaucracy to evict his people from their traditional homeland, was chosen by IUCN to host the largest conservation convention in history.

The response of big conservation, in Bangkok and elsewhere, has been to deny that they are party to the evictions while generating reams of promotional material about their affection for and close relationships with indigenous peoples. "We recognize that indigenous people have perhaps the deepest understanding of the earth's living resources," says the Conservation International chairman and CEO, Peter Seligman, adding that "we firmly believe that indigenous people must have ownership, control, and title of their lands." Such messages are carefully projected toward major funders of conservation, which in response to the aforementioned Ford Foundation reports and other press have become increasingly sensitive to indigenous peoples and their struggles for cultural survival.

Financial support for international conservation has in recent years expanded well beyond the individuals and family foundations that seeded the movement to include very large foundations like the Ford, MacArthur, and Gordon and Betty Moore foundations, as well as the World Bank, its Global Environment Facility, foreign governments, USAID, a host of bilateral and multilateral banks, and transnational corporations. During the 1990s USAID alone pumped almost $300 million into the international conservation movement, which it had come to regard as a vital adjunct to economic prosperity. The five largest conservation organizations, CI, TNC, and WWF among them, absorbed over 70 percent of that expenditure. Indigenous communities received none of it. The Moore Foundation made a singular ten-year commitment of nearly

$280 million, the largest environmental grant in history, to just one organization — Conservation International. And all of the BINGOs have become increasingly corporate in recent years, both in orientation and affiliation. The Nature Conservancy now boasts almost two thousand corporate sponsors, while Conservation International has received about $9 million from its two hundred fifty corporate "partners."

With that kind of financial and political leverage, and with chapters in almost every country of the world, millions of loyal members, and nine-figure budgets, CI, WWF, and TNC have undertaken a hugely expanded global push to increase the number of so-called protected areas (PAs) — parks, reserves, wildlife sanctuaries, and corridors created to preserve biological diversity. In 1962 there were some 1,000 official PAs worldwide. Today there are 108,000, with more being added every day. The total area of land now under conservation protection worldwide has doubled since 1990, when the World Parks Commission set a goal of protecting 10 percent of the planet's surface. That goal has been exceeded, as over 12 percent of all land, a total area of 11.75 million square miles, is now protected. That's an area greater than the entire land mass of Africa.

At first glance, so much protected land seems undeniably positive, an enormous achievement of very good people doing the right thing for our planet. But the record is less impressive when the impact upon native people is considered. For example, during the 1990s the African nation of Chad increased the amount of national land under protection from 0.1 to 9.1 percent. All of that land had been previously inhabited by what are now an estimated six hundred thousand conservation refugees. No other country besides India, which officially admits to 1.6 million, is even counting this growing new class of refugees. World estimates offered by the UN, IUCN, and a few anthropologists range from 5 million to tens of millions of people. Charles Geisler, a sociologist at Cornell University who has studied displacements in Africa, is certain the number on that continent alone exceeds 14 million.

The true worldwide figure, if it were ever known, would depend upon the semantics of words like "eviction," "displacement," and "refugee," over which parties on all sides of the issue argue endlessly. The larger point is that conservation refugees exist on every continent but Antarctica, and by most accounts live far more difficult lives than they once did, banished from lands they thrived on for hundreds, even thousands of years.

The practice of removing people from protected areas began in the United States in 1864 with the military expulsion of Miwok and Ahwahnee Indians from their four-thousand-year-old settlements in Yosemite Valley. During the California Gold Rush the valley and its native communities had been "discovered" by white settlers. One of them, a miner and wilderness romantic named Lafayette Burnell, swooned over the lush beauty of the valley as he watched James Savage, commander of the Mariposa Battalion, burn Indian villages and acorn caches to the ground — a first step to starving and freezing the Miwok into submission.

Burnell approved of the torching. Fancying himself a passionate conservationist, he was determined to "sweep the territory of any scattered bands that might infest it." And swept it was, a process that lasted until 1969, when the last Miwok village was evacuated from the national park. Similar treatment was experienced by the Shoshone, Lakota, Bannock, Crow, Nez Perce, Flathead, and Blackfeet, all of whom at one time or another occupied and hunted in what is now Yellowstone National Park. And many parks have followed, followed by many evictions.

John Muir, a forefather of the American conservation movement, argued that "wilderness" should be cleared of all inhabitants and set aside to satisfy the urbane human's need for recreation and spiritual renewal. It was a sentiment that became national policy with the passage of the 1964 Wilderness Act, which defined wilderness as a place "where man himself is a visitor who does not remain." One should not be surprised to find hardy residues of these sentiments among traditional conservation groups. The preference for "virgin" wilderness has lingered on in a movement that has tended to value all nature but human nature and refused to recognize the positive wildness in human beings.

Expulsions continue around the world to this day, albeit under less violent circumstances than the atrocious Miwok massacres. The Indian government, which evicted one hundred thousand *adivasis* (rural peoples) in Assam between April and July of 2002, estimates that another two or three million more will be displaced over the next decade. The policy is largely in response to a 1993 lawsuit brought by WWF, which demanded that the government increase PAs by 8 percent, mostly in order to protect tiger habitat. A more immediate threat involves the impending removal of several Mayan communities from the Montes Azules region of Chiapas, Mexico, a process, begun in the mid-1970s with the intent to preserve virgin tropical forest, which could still quite easily spark a civil war. Conservation International is deeply immersed in that controversy, as are a host of extractive industries.

Tensions are also high in the Enoosupukia region of Kenya, where two thousand members of the ancient Ogiek community were recently ordered to leave the Mau forest, where they have thrived as hunter-gatherers for centuries. ANY PERSON FOUND INSIDE THE TRUST LAND AREA SHALL BE EVICTED/ARRESTED reads a warning posted by the government. Once the Ogiek villages were cleared, all structures were burned to the ground. And while the stated intent of the Kenyan government is environmental, the Ogiek note that their land has been deeded over to powerful members of former president Daniel Arap Moi's Kalenjin tribe and that vast regions of the forest are being clearcut. An Ogiek defender, Kenya's deputy minister of environment, Wangari Maathai, recently told *Newsweek* that "the Ogiek are a class of their own." A lifelong environmental activist and Nobel laureate, Maathai remembers what the Mau Forest was like when the Ogiek lived there: no roads, no logging, plenty of biodiversity. The World Wildlife Fund, which is active in the area, has been careful not to promote the Ogiek eviction but has also done nothing to stop it.

Meanwhile, over the past decade each of the BINGOs and most of the international agencies they work with have issued formal and heartfelt declarations in support of indigenous peoples and their territorial rights. The Nature Conservancy's "Commitment to People" statement declares, "We respect the needs of local communities by developing ways to conserve biological diversity while at the same time enabling humans to live productively and sustainably on the landscape." After endorsing the UN's draft declaration on the rights of indigenous peoples, WWF adopted its own statement of principles upholding the rights of indigenous peoples to own, manage, and control their lands and territories — a radical notion for many governments. In 1999 the World Commission on

Protected Areas formally recognized indigenous peoples' rights to "sustainable, traditional use" of their lands and territories.

The following year the IUCN adopted a bold set of principles for establishing protected areas, which states unequivocally, "The establishment of new protected areas on indigenous and other traditional peoples' . . . domains should be based on the legal recognition of collective rights of communities living within them to the lands, territories, waters, coastal seas, and other resources they traditionally own or otherwise occupy or use."

Tribal people, who tend to think and plan in generations, rather than weeks, months, and years, are still waiting to be paid the consideration promised in these thoughtful pronouncements. Of course the UN draft declaration is the prize because it must be ratified by so many nations. The declaration has failed to pass mainly because powerful leaders such as Tony Blair and George Bush threaten to veto it, arguing that there is not and should never be such a thing as collective human rights.

Sadly, the human rights and global conservation communities remain at serious odds over the question of displacement, each side blaming the other for the particular crisis they perceive. Conservation biologists argue that by allowing native populations to grow, hunt, and gather in protected areas, anthropologists, cultural preservationists, and other supporters of indigenous rights become complicit in the decline of biological diversity. Some, like the Wildlife Conservation Society's outspoken president, Steven Sanderson, believe that the entire global conservation agenda has been "hijacked" by advocates for indigenous peoples, placing wildlife and biodiversity in peril. "Forest peoples and their representatives may speak for the forest," Sanderson has said. "They may speak for their version of the forest; but they do not speak for the forest we want to conserve." WCS, originally the New York Zoological Society, is a BINGO lesser in size and stature than the likes of TNC and CI, but more insistent than its colleagues that indigenous territorial rights, while a valid social issue, should be of no concern to wildlife conservationists.

Human rights groups, such as Cultural Survival, First Peoples Worldwide, EarthRights International, Survival International, and the Forest Peoples Programme argue the opposite, accusing some of the BINGOs and governments like Uganda's of destroying indigenous cultures, the diversity of which they deem essential to the preservation of biological diversity.

One attempt to bridge this unfortunate divide is the "market-based solution." BINGOs endorse ecotourism, bioprospecting, extractive reserves, and industrial partnerships that involve such activities as building nature resorts, leading pharmaceutical scientists to medicinal plants, gathering nuts for Ben and Jerry's ice cream, or harvesting plant oils for the Body Shop as the best way to protect both land and community with a single program. Global conservation Web sites and annual reports feature stunning photographs of native people leading nature tours and harvesting fair-trade coffee, Brazil nuts, and medicinal plants. But no native names or faces can be found on the boards of the BINGOs that promote these arrangements.

Market-based solutions, which may have been implemented with the best of social and ecological intentions, share a lamentable outcome, barely discernible behind a smoke screen of slick promotion. In almost every case indigenous people are moved into the money economy without the means to participate in it fully. They become permanently indentured as park rangers (never wardens), porters, waiters, harvesters, or, if they manage to learn a European language, ecotour guides. Under this model, "conservation" edges ever closer to "development," while native communities are assimilated into the lowest ranks of national cultures.

Given this history, it should be no surprise that tribal peoples regard conservationists as just another colonizer — an extension of the deadening forces of economic and cultural hegemony. Whole societies like the Batwa, the Maasai, the Ashinika of Peru, the Gwi and Gana Bushmen of Botswana, the Karen and Hmong of Southeast Asia, and the Huarani of Ecuador are being transformed from independent and self-sustaining into deeply dependent and poor communities.

When I traveled throughout Mesoamerica and the Andean Amazon watershed last fall visiting staff members of CI, TNC, WCS, and WWF, I was looking for signs that an awakening was on the horizon. The field staff I met were acutely aware that the spirit of exclusion survives in the headquarters of their organizations, alongside a subtle but real prejudice against "unscientific" native wisdom. Dan

Campbell, TNC's director in Belize, conceded, "We have an organization that sometimes tries to employ models that don't fit the culture of nations where we work." And Joy Grant, in the same office, said that as a consequence of a protracted disagreement with the indigenous peoples of Belize, local people "are now the key to everything we do."

"We are arrogant" was the confession of a CI executive working in South America, who asked me not to identify her. I was heartened by her admission until she went on to suggest that this was merely a minor character flaw. In fact, arrogance was cited by almost all of the nearly one hundred indigenous leaders I met with as a major impediment to constructive communication with big conservation.

Luis Suarez, the new director of CI in Ecuador, seems to be aware of that. "Yes," he said, "CI has made some serious blunders with indigenous organizations within the past four years, not only in Ecuador but also in Peru." And he admitted to me that his organization was at that very moment making new enemies in Guyana, where CI had worked with the Wai Wai peoples on the establishment of a protected area but had simultaneously ignored another tribe, the Wapishana, whose six communities will be encompassed by the park.

If field observations and field workers' sentiments trickle up to the headquarters of CI and the other BINGOs, there could be a happy ending to this story. There are already positive working models of socially sensitive conservation on every continent, particularly in Australia, Bolivia, Nepal, and Canada, where national laws that protect native land rights leave foreign conservationists no choice but to join hands with indigenous communities and work out creative ways to protect wildlife habitat and sustain biodiversity while allowing indigenous citizens to thrive in their traditional settlements.

However, in most such cases it is the native people who initiate the creation of a reserve, which is more likely to be called an "indigenous protected area" (IPA) or a "community conservation area" (CCA). IPAs are an invention of Australian aboriginals, many of whom have regained ownership and territorial autonomy under new treaties with the national government, and CCAs are appearing around the world, from Lao fishing villages along the Mekong

River to the Mataven Forest in Colombia, where six indigenous tribes live in 152 villages bordering a four-million-acre ecologically intact reserve. The tribes manage a national park within the reserve and collectively own considerable acreage along its border. Before the Mataven conservation area was created, the indigenous communities mapped the boundaries of the land to be protected, proposed their own operating rules and restrictions, and sought independent funding to pay for management of the reserve, which today is regarded worldwide as a model of indigenous conservation.

The Kayapo, a nation of Amazonian Indians with whom the Brazilian government and CI have formed a cooperative conservation project, is another such example. Kayapo leaders, renowned for their militancy, openly refused to be treated like just another stakeholder in a two-way deal between a national government and a conservation NGO, as is so often the case with cooperative management plans. Throughout negotiations they insisted upon being an equal player at the table, with equal rights and land sovereignty. As a consequence, the Xingu National Park, the continent's first Indian-owned park, was created to protect the lifeways of the Kayapo and other indigenous Amazonians who are determined to remain within the park's boundaries.

In many locations, once a CCA is established and territorial rights are assured, the founding community invites a BINGO to send its ecologists and wildlife biologists to share in the task of protecting biodiversity by combining Western scientific methodology with indigenous ecological knowledge. And on occasion the founders will ask for help in negotiating with reluctant governments. For example, the Guarani Izoceños people in Bolivia invited the Wildlife Conservation Society to mediate a comanagement agreement with their government, which today allows the tribe to manage and own part of the new Kaa-Iya del Gran Chaco National Park.

Too much hope should probably not be placed in a handful of successful comanagement models or a few field staffers' epiphanies. The unrestrained corporate lust for energy, hardwood, medicines, and strategic metals is still a considerable threat to indigenous communities, arguably a larger threat than conservation. However, the lines between the two are being blurred. Particularly problematic is the fact that international conservation organizations remain comfortable working in close quarters with some of the most aggressive global resource prospectors, such as Boise Cascade, Chevron-Texaco, Mitsubishi, Conoco-Phillips, International Paper, Rio Tinto Mining, Shell, and Weyerhaeuser, all of whom are members of a CI-created entity called the Center for Environmental Leadership in Business. Of course if the BINGOs were to renounce their corporate partners, they would forfeit millions of dollars in revenue and access to global power, without which they sincerely believe they could not be effective.

And there are some respected and influential conservation biologists who still strongly support top-down, centralized "fortress" conservation. Duke University's John Terborgh, for example, author of the classic *Requiem for Nature,* believes that comanagement projects and CCAs are a huge mistake. "My feeling is that a park should be a park, and it shouldn't have any resident people in it," he says. He bases his argument on three decades of research in Peru's Manu National Park, where native Machiguenga Indians fish and hunt animals with traditional weapons. Terborgh is concerned that they will acquire the motorboats, guns, and chain saws used by their fellow tribesmen outside the park, and that biodiversity will suffer. Then there's paleontologist Richard Leakey, who at the 2003 World Parks Congress in South Africa set off a firestorm of protest by denying the very existence of indigenous peoples in Kenya, his homeland, and arguing that "the global interest in biodiversity might sometimes trump the rights of local people."

Not all of Leakey's colleagues agree with him. Many conservationists are beginning to realize that most of the areas they have sought to protect are rich in biodiversity precisely because the people who were living there had come to understand the value and mechanisms of biological diversity. Some will even admit that wrecking the lives of 10 million or more poor, powerless people has been an enormous mistake — not only a moral, social, philosophical, and economic mistake, but an ecological one as well. Others have learned from bitter experience that national parks and protected areas surrounded by angry, hungry people who describe themselves as "enemies of conservation" are generally doomed to fail. As Cristina Eghenter of WWF observed after working with communities surrounding the Kayan Mentarang National Park in Borneo, "It is becoming increasingly evident that conservation objectives can rarely be obtained or sustained by imposing policies that produce negative impacts on indigenous peoples."

More and more conservationists seem to be wondering how, after setting aside a "protected" land mass the size of Africa, global biodiversity continues to decline: might there be something terribly wrong with this plan? This question is particularly apt since the Convention on Biological Diversity has documented the astounding fact that in Africa, where so many parks and reserves have been created and where indigenous evictions run highest, 90 percent of biodiversity lies outside of protected areas. If we want to preserve biodiversity in the far reaches of the globe, places that are in many cases still occupied by indigenous people living in ways that are ecologically sustainable, history is showing us that the dumbest thing we can do is kick them out.

JOHN HOCKENBERRY

The Blogs of War

FROM *Wired*

THE SNAPSHOTS OF IRAQI PRISONERS being abused at Abu Ghraib were taken by soldiers and shared in the digital military netherworld of Iraq. Their release to the world in May 2004 detonated a media explosion that rocked a presidential campaign, cratered America's moral high ground, and demonstrated how even a superpower could be blitzkrieged by some homemade downloadable porn. In the middle of it all, a lone reservist sergeant stationed on the Iraqi border posed a simple question: "I cannot help but wonder upon reflection of the circumstances, how much longer we will be able to carry with us our digital cameras, or take photographs and document the experiences we have had."

The writer was twenty-four-year-old Chris Missick, a soldier with the army's 319th Signal Battalion and author of the blog "A Line in the Sand." While balloon-faced cable pundits shrieked about the scandal, Missick was posting late at night in his army-issue "blacks," with a mug of coffee and a small French press beside him, his laptop blasting Elliott Smith's "Cupid's Trick" into his headphones. He quickly seized on perhaps the most profound and crucial implication of Abu Ghraib:

> Never before has a war been so immediately documented, never before have sentiments from the front scurried their way to the home front with such ease and precision. Here I sit, in the desert, staring daily at the electric fence, the deep trenches, and the concertina wire that separates the border of Iraq and Kuwait, and write home and upload my daily reflections and opinions on the war and my circumstances here, as well as some of the pictures I have taken along the way. It is amazing, and em-

> powering, and yet the question remains, should I as a lower enlisted soldier have such power to express my opinion and broadcast to the world a singular soldier's point of view? To those outside the uniform who have never lived the military life, the question may seem absurd, and yet, as an example of what exists even in the small following of readers I have here, the implications of thought expressed by soldiers daily could be explosive.

His sober assessments of the potential of free speech in a war zone began attracting a wider following, eventually logging somewhere north of 100,000 pageviews. No blogging record, but rivaling the wonkish audience for the Pentagon's daily briefing on C-Span or Department of Defense (DOD) press releases.

Missick is just one voice — and a very pro-Pentagon one at that — in an oddball online Greek chorus narrating the conflict in Iraq. It includes a core group of about one hundred regulars and hundreds more loosely organized activists, angry contrarians, jolly testosterone fuckups, self-appointed pundits, and would-be poets who call themselves milbloggers, as in military bloggers. Whether posting from inside Iraq on active duty, from noncombat bases around the world, or even from their neighborhoods back home after being discharged — where they can still follow events closely and deliver their often blunt opinions — milbloggers offer an unprecedented real-time, real-life window on war and the people who wage it. Their collective voice competes with and occasionally undermines the DOD's elaborate message machine and the much-loathed mainstream media, usually dismissed as MSM.

Milbloggers constitute a rich subculture with a refreshing candor about the war, expressing views ranging from far right to far left. They also offer helpful tips about tearing down an M-16, recipes for beef stew (hint: lots of red wine), reviews of the latest episode of *24,* extremely technical discussions of Humvee armor configurations, and exceptionally raw accounts of field-hospital chaos, gore, and heroism.

For now, the Pentagon officially tolerates this free-form online journalism and in-house peanut gallery, even as the brass takes cautious steps to control it. A new policy instituted this spring requires all military bloggers inside Iraq to register with their units. It directs commanders to conduct quarterly reviews to make sure bloggers aren't giving out casualty information or violating operational security or privacy rules. Commanding officers shut down a blog that reported on the medical response to a suicide bombing late last year in Mosul. The army has also created the Army Web Risk Assessment Cell to monitor compliance. And *Wired* has learned that a Pentagon review is under way to better understand the overall implications of blogging and other Internet communications in combat zones.

"It's a new world out there," says Christopher Conway, a lieutenant colonel and DOD spokesperson. "Before, you would have to shake down your soldiers for matches that might light up and betray a position. Today, every soldier has a cell phone, beeper, game device, or laptop, any one of which could pop off without warning. Blogging is just one piece of the puzzle."

Strong opinions throughout the military ranks, in and out of wartime, are nothing new. But online technology in the combat zone has suddenly given those opinions a mass audience and an instantaneous forum for the first time in the history of warfare. On the twenty-first-century battlefield, the campfire glow comes from a laptop computer, and it's visible around the world.

"In World War II, letters basically didn't arrive for months," says Michael Bautista, an Idaho National Guard corporal based in Kirkuk whose grandfather served in World War II and who blogs as Ma Deuce Gunner (named for the trusty M-2 machine gun he calls Mama). "What I'm doing and what my fellow bloggers are doing is groundbreaking."

If you're stuck in southern Baghdad in the dusty gray fortress called Camp Falcon and find yourself in need of 50-caliber machine-gun ammo, chopper fuel, toilet paper, or M&M's, you call Danjel Bout, a thirty-two-year-old captain and logistics officer from the California National Guard who blogs as Thunder 6. He's been stationed here with the army's 3rd Infantry Division for most of 2005. When he's not chasing down requisitions of supplies or out on patrol hunting insurgents, Bout is posting about the details of army life in language evocative of literary warbloggers of yore like Thucydides, Homer, Thomas Paine, and John Donne.

> Sleep, blessed, blissful, wonderful sleep. Mother's milk. A full harvest in a time of famine. The storm that breaks the drought. It is the drug of

> choice here — assiduously avoided because of the never-ending chain of missions, but always craved. If rarity is the measure of a substance's worth, then here in Iraq, sleep carries a price beyond words. There is no more precious moment in my day than the sublime instant where my mind flickers between consciousness and the dreamworld. In that sliver of time the day seems to shimmer and melt like one of Dalí's paintings — leaving only honey sweet dreams of my other life far from Arabia.

Bout's blog, "365 and a Wakeup," is unlikely to put you to sleep. It's one of the most genuine accounts anywhere of what life is like for a soldier in Iraq. The captain can be spotted composing and editing his posts on his laptop from the roof of one of Camp Falcon's dusty buildings in the dark early-morning hours, or in a scarce patch of shade during a rare moment of daylight downtime. His posts are sharply rendered parables and small, often powerful scenes built on details of the violent world around him. "I just kind of bookmark the things I see during the day so I can reflect on them later. There's almost nothing about life here that isn't interesting in some way."

Thunder 6 is the oldest of eight siblings in a devout Catholic family. His dad is a computer technician, his mom a horticulture therapist. This former altar boy and longtime reservist left the touchy-feely psychology Ph.D. program at the University of California, Davis, after September 11, grabbed an M-16 rifle and a Beretta 9-mm sidearm, and went all-infantry. Trained as an army ranger, he saw action in Kuwait and Bosnia and claims to have no yearning for his former scholarly life. "I was coasting through college," he says, "and the army spoke to honor and camaraderie and things I really believed in."

While Bout's blog is all about his emotional connection to the army and very little about the daily bang-bang of Iraq, there are lots of milbloggers who will take you straight to the front lines, posting first-person accounts of the fighting and beating some newspaper reports of the same battle filed by embedded journalists. By the crude light of a small bulb and the backlit screen of his Dell laptop, Neil Prakash, a first lieutenant, posted some of the best descriptions of the fighting in Fallujah and Baquba last fall:

> Terrorists in headwraps stood anywhere from 30 to 400 meters in front of my tank. They stopped, squared their shoulders at us just like in an

> old-fashioned duel, and fired RPGs at our tanks. So far there hadn't been a single civilian in Task Force 2–2 sector. We had been free to light up the insurgents as we saw them. And because of that freedom, we were able to use the main gun with less restriction.

Prakash was awarded the Silver Star this year for saving his entire tank task force during an assault on insurgents in Iraq's harrowing Sunni Triangle. He goes by the handle Red 6 and is the author of "Armor Geddon." For him, the poetry of warfare is in the sounds of exploding weapons and the chaos of battle.

"It's mind-blowing what this stuff can do," Prakash tells me by phone from Germany, where his unit moved after rotating out of Iraq earlier this year. One of his favorite sounds is that of an F16 fighter on a strafing run. "It's like a cat in a blender ripping the sky open — if the sky was made out of a phone book." He is from India, the land of Gandhi, but he loves to talk about blowing things up. "It's just sick how badass a tank looks when it's killing."

Prakash is the son of two upstate New York dentists and has a degree in neuroscience from Johns Hopkins. He's a naturalized American citizen, born near Bangalore, and he describes growing up in the United States and his decision to join the military as something like *Bend It Like Beckham* meets *The Terminator.* He says he admired the army's discipline and loved the idea of driving a tank. He knew that if he didn't join the army, he might end up in medical school or some windowless office in a high-tech company. With a bit of bluster, Prakash claims that for him, the latter would be more of a nightmare scenario than ending up in the line of fire of insurgents. "It was a choice between commanding the best bunch of guys in the world and being in a cubicle at Dell Computer in Bangalore right now helping people from Bum-fuck, USA, format their hard drives."

It's taken some adjustment, but Prakash says his parents basically support his army career, although his father can't conceal his anxiety about having a son in Iraq. Prakash says he blogs to assure the folks back home that he's safe, to let his friends all over the world know what's going on, and to juice up the morale in his unit. "The guys get really excited when I mention them."

By the time Prakash left Iraq early this year, the readers of "Armor Geddon" extended far beyond family and friends. He still posts from his base in Germany and is slowly trying to complete a blog memoir of his and his fellow soldiers' experiences in the battle for Fallujah.

The most widely read milbloggers engage in the twenty-first-century contact sport called punditry and, like their civilian counterparts, follow few rules of engagement. They mobilize sympathizers to ship body armor to reserve units in combat, raise funds for families of wounded soldiers, deliver shoes to barefoot Afghani kids, and even take aim at media big shots. It was milblogger pundits who helped bring down Eason Jordan, a senior executive at CNN who resigned earlier this year over remarks he made that U.S. troops were targeting reporters in Iraq.

One important milblogger who weighed in on the Jordan affair is a secretive twenty-year career army GI who goes by the handle Greyhawk. His blog, "The Mudville Gazette," investigated the incident and concluded that Iraq-based reporters disputed Jordan's claim. He's unhappy that a more thorough news investigation wasn't conducted. Other bloggers call Greyhawk "the father of us all" and credit him with coining the term *milblogger* shortly after he started "Mudville" in March 2003. In an e-mail interview — Greyhawk wouldn't agree to "voice-com" or a "face-to face" — he writes proudly of his lifetime pageviews, which recently exceeded 1.7 million (700,000 of those have come in 2005): "Mudville is far and away the largest, oldest, widest-read active-duty MilBlog in the World. It's all in how you make the words line up and dance."

Then there's Blackfive: "I'm just a guy with a blog and I know how to use it," says this modest former army intelligence officer and paratrooper, who gives his real name only as Matt. He prefers the nom de guerre of his popular site. His peers voted "Blackfive" the best military blog in the 2004 Weblog Awards, beating out such contenders as "Froggy Ruminations," "The Mudville Gazette," "2Slick," and "My War." "Blackfive" is a popular forum for analysis of the war and strident, argumentative warnings about media bias. It's nearly as cluttered with ads as "The Drudge Report," and the sales pitches mostly hawk "liberal-baiting merchandise." There are pictures of attractive women holding high-powered weapons, dozens of links to conservative books and films, and even the occasional big spender like Amazon.com. Blackfive also sells his own T-shirts to benefit military charities.

He says that milblogging is the result of an explosion of communications technology throughout the military and an increase in brainpower among the lower ranks. "The educational level of sergeants and below is out of control." Blackfive himself has degrees in archaeology and computer science and avidly follows the postings of fellow bloggers. He describes Neil Prakash as "borderline Einstein" and Danjel Bout as "a real rock star." In his last deployment, Blackfive's unit had two such brainiacs, a sergeant with an M.B.A. and another with a master's in economics from the University of Chicago.

Blackfive is retired now, honorably discharged and working as an IT executive for a big civilian company. He blogs from Chicago and confidently claims he can mobilize thousands of people and their wallets, all from a wireless hot spot at his local Starbucks. He stays in the shadows because he believes that his company would not approve of his blog or of his unabashed support for the U.S. war.

The site has become a destination for thousands of information junkies and influential opinion makers. According to "TruthLaid Bear," which tracks blog traffic for advertisers, "Blackfive" is regularly in the top one hundred blogs and averages five thousand unique visits a day. During the height of the war, traffic to "Blackfive" spiked when some high-profile conservatives linked to the site.

"My brother followed a link from *National Review* to me, and somebody, I think it was Jonah Goldberg" — a somebody who is only the editor of *National Review* — "told him that four or five of the biggest think tanks read my blog every day."

Goldberg confirms that at times he turns to military blogs to supplement and sometimes contradict information coming out of traditional media sources. "'Blackfive' was good, and in the blog world if you offer something unique, you make eyeballs sticky."

Since World War I the military has opened the letters soldiers sent back home from the battlefield and sometimes censored the dispatches of war correspondents. Now mail leaves the battlefield already open to the world. Anyone can publicly post a dispatch, and if the Pentagon reads these accounts at all, it's at the same time as the rest of us. The new policy requiring milbloggers to register their sites does not apply to soldiers outside Iraq, but nearly all of the bloggers contacted for this article say that the current system of few restrictions can't possibly last. Blackfive and Greyhawk wonder what the landscape will look like after the Pentagon finishes its review of global digital security. So far, the DOD is giving no hints.

Michael Cohen, a major and doctor with the 67th Combat Support Hospital based in Mosul, touched a nerve at the Pentagon late last year with his blog, "67cshdocs." Before he began posting, Cohen turned himself into a local private broadband provider in order to set up his own network outside the one provided to the field hospital. "Some of the docs suggested that life would be really good if we could get Internet into our nice trailers."

Cohen bought his network setup online and had it shipped directly to him in Mosul. For the oversize satellite dish, he had to get creative. He ordered it from Bentley Walker, a satellite broadband service provider, and they sent it to his wife's house in Germany. On a medical escort flight to Germany for a wounded soldier, Cohen persuaded the air force to let him hand-carry the dish onto a transport for the return trip. After about six weeks of agonized troubleshooting on a hot rooftop, the network was up and running. "We had pretty decent bandwidth," he says, "two meg downlink and one meg up. It was better than the hospital."

Cohen says the system supported webcams linking people back home, its own instant messaging system, live gaming, and, he theorizes, a robust trade in porn. "If you were to make the series *M*A*S*H* about today's army, Radar would be an IT guy and he'd be more popular than Hawkeye."

Then Cohen started to blog on his homegrown network. Originally it was an attempt to stay in touch with family and friends, but when a suicide bomber killed twenty-two people last December in a mess tent, Cohen began detailing how doctors dealt with the carnage. His moving account drew attention from worldwide press as well as parents desperate to know the fate of their loved ones:

> The lab was running tests and doing a blood drive to collect more blood. The pharmacy was preparing intravenous medications and drips like crazy. Radiology was shooting plain films and CT scans like nobody's business. We were washing out wounds, removing shrapnel, and casting fractures. We put in a bunch of chest tubes. Because of all the patients on suction machines and mechanical ventilators, the noise in the ICU was so loud everyone was screaming at each other just to communicate.

Here are some of our statistics. They are really quite amazing: 91 total patients arrived.

18 were dead on arrival.

4 patients died of wounds shortly after arrival — all of these patients had non-survivable wounds.

Of the 69 remaining patients, 20 were transferred to military hospitals in other locations in Iraq.

This left 49 patients for us to treat and disposition.

Cohen posted mesmerizing details about the medical hardware and surgical procedures used to save lives on that bloody day. And then, without warning, it was over.

"My doctor boss came to me and said, hey, we need to talk. There are some people in the chain of command who believe there are things in your blog that violate army regulations." Cohen was shocked. He hadn't used names or talked about military operations. But his impression was that the information he provided about medical capability in the field worried senior officers at Central Command. At first the army asked Cohen to shut down his entire satellite network, which at its peak was serving forty-two families, but ultimately decided against it.

"I think they didn't want a hornet's nest," Cohen says. Instead, Cohen stopped blogging.

Back in Germany now, where he says he spends more time delivering the latest R-and-R babies than treating battlefield casualties, Cohen says that he was tempted to challenge the shutdown, but since he was close to going home anyway, he went along with the decision. The Pentagon will not comment specifically on Cohen's situation except to reiterate its policy that blogs should not reveal any casualty information that could upset next of kin or any details that might jeopardize operational security.

Army reservist Jason Hartley's popular and notoriously irreverent blog, "Just Another Soldier," also provoked the higher-ups; last summer, his commanding officer ordered him to shut it down. Hartley wrote with a fuck-you swagger that may partly explain why he's not blogging anymore: "Being a soldier is to live in a world of shit. From the pogues who cook my food and do my laundry to the Apache pilots and the Green Berets who do all the Hollywood stuff, our lives are in a constant state of suck."

Hartley got a lot of mileage out of a post about a soldier who was assembling a rifle blindfolded. Another soldier in his unit, as a joke, handed the assembler a certain piece of his anatomy instead of the tool he asked for.

"I told the story and asked the question, Who is more gay, the guy who touches a dick, or someone who allows a soldier to touch his dick?" This pressing infantry-level controversy hit a chord with the über-blogger and noted pundit-of-all-things-queer Andrew Sullivan. "Sullivan was kind and wrote that he liked my site," Hartley recalls.

The Pentagon won't say why, but it ordered Hartley to shut down his blog. He did for a while. Then he resumed blogging a few months later, without asking permission, and was busted for defying a direct order and demoted from sergeant to specialist. He chose not to file an appeal and has returned to civilian life, though he's still in the reserves. His memoir about his time in Iraq will be published by HarperCollins.

If you read "A Line in the Sand," it's hard to imagine Chris Missick offending Pentagon brass. He is careful not to criticize his superiors and will tell you he has aspirations to run for Congress. While waiting for an early-morning plane to take him back home to southern California, Missick confesses that his biggest blog-related scandal is a romantic one. His stateside girlfriend when he left for Iraq was displaced by another woman, someone Missick says fell in love with him by reading his blog. "When I get home I kinda need to sort that out." (He kinda did and now has yet another girlfriend. Let's hope she likes Elliott Smith's music.)

Prakash remains in Germany, awaiting orders to jump back into his beloved tank, which he calls Ol' Blinky. He says he has no plans to resume his study of neuroscience, although it wasn't completely useless in Iraq. "Neuroscience actually came in handy when I had to explain to my guys exactly why doing Ecstasy in a tank when it's 140 degrees out on a road that's blowing up every day is a really bad idea."

Danjel Bout, aka Thunder 6, is looking to get home safely, keeping his head down on the streets of southern Baghdad and in his blog. He says the real value of milblogging may be that it brings to the United States the reality of what is becoming a long war. "I don't purposely leave out the moments when our bodies hit the adrenal dump switch, I just don't focus exclusively on them." More typical are his vignettes of Iraqi civilians interacting with U.S. soldiers, or the sad tale of the death of a guardsman who had the chance to go home and instead requested another tour of duty, only to be killed by an improvised explosive device.

"Americans are raised on a steady diet of action films and sound bites that slip from one supercharged scene to another," he says, "leaving out all the confusing decisions and subtle details where most people actually spend their lives. While that makes for a great story, it doesn't reveal anything of lasting value. For people to really understand our day-to-day experience here, they need more than the highlights reel. They need to see the world through our eyes for a few minutes."

Which suggests, at the very least, that this UC Davis psych-major dropout turned milblogger was perhaps paying more attention in class than he lets on.

JOHN HORGAN

The Forgotten Era of Brain Chips

FROM *Scientific American*

IN THE EARLY 1970s José Manuel Rodriguez Delgado, a professor of physiology at Yale University, was among the world's most acclaimed — and controversial — neuroscientists. In 1970 the *New York Times Magazine* hailed him in a cover story as the "impassioned prophet of a new 'psychocivilized society' whose members would influence and alter their own mental functions." The article added, though, that some of Delgado's Yale colleagues saw "frightening potentials" in his work.

Delgado, after all, had pioneered that most unnerving of technologies, the brain chip — an electronic device that can manipulate the mind by receiving signals from and transmitting them to neurons. Long the McGuffins of science fiction, from *The Terminal Man* to *The Matrix,* brain chips are now being used or tested as treatments for epilepsy, Parkinson's disease, paralysis, blindness, and other disorders. Decades ago Delgado carried out experiments that were more dramatic in some respects than anything being done today. He implanted radio-equipped electrode arrays, which he called "stimoceivers," in cats, monkeys, chimpanzees, gibbons, bulls, and even humans, and he showed that he could control subjects' minds and bodies with the push of a button.

Yet after Delgado moved to Spain in 1974, his reputation in the United States faded, not only from public memory but from the minds and citation lists of other scientists. He described his results in more than five hundred peer-reviewed papers and in a widely reviewed 1969 book, but these are seldom cited by modern researchers. In fact, some familiar with his early work assume that he died.

But Delgado, who recently moved with his wife, Caroline, from Spain to San Diego, is very much alive and well, and he has a unique perspective on modern efforts to treat various disorders by stimulating specific areas of the brain.

When Lobotomies Were the Rage

Born in 1915 in Ronda, Spain, Delgado went on to earn a medical degree from the University of Madrid in the 1930s. Although he has long been dogged by rumors that he supported the fascist regime of Francisco Franco, he actually served in the medical corps of the Republican army (which opposed Franco during Spain's civil war) while he was a medical student. After Franco crushed the Republicans, Delgado was detained in a concentration camp for five months before resuming his studies.

He originally intended to become an eye doctor, like his father. But a stint in a physiology laboratory — plus exposure to the writings of the great Spanish neuroscientist Santiago Ramón y Cajal — left him entranced by "the many mysteries of the brain. How little was known then. How little is known now!" Delgado was particularly intrigued by the experiments of the Swiss physiologist Walter Rudolf Hess. Beginning in the 1920s, Hess had demonstrated that he could elicit behaviors such as rage, hunger, and sleepiness in cats by electrically stimulating different spots in their brains with wires.

In 1946 Delgado won a yearlong fellowship at Yale. In 1950 he accepted a position in its department of physiology, then headed by John Fulton, who played a crucial role in the history of psychiatry. In a 1935 lecture in London, Fulton had reported that a violent, "neurotic" chimpanzee named Becky had become calm and compliant after surgical destruction of her prefrontal lobes. In the audience was a Portuguese psychiatrist, Egas Moniz, who started performing lobotomies on psychotic patients and claimed excellent results. After Moniz won a Nobel Prize in 1949, lobotomies became an increasingly popular treatment for mental illness.

Initially disturbed that his method of pacifying a chimpanzee had been applied to humans, Fulton later became a cautious proponent of psychosurgery. Delgado disagreed with his mentor's stance. "I thought Fulton and Moniz's idea of destroying the brain was absolutely horrendous," Delgado recalls. He felt it would be "far more conservative" to treat mental illness by applying the electrical stimulation methods pioneered by Hess — who shared the 1949 prize with Moniz. "My idea was to *avoid* lobotomy," Delgado says, "with the help of electrodes implanted in the brain."

One key to Delgado's scientific success was his skill as an inventor; a Yale colleague once called him a "technological wizard." In his early experiments, wires ran from implanted electrodes out through the skull and skin to bulky electronic devices that recorded data and delivered electrical pulses. This setup restricted subjects' movements and left them prone to infections. Hence, Delgado designed radio-equipped stimoceivers as small as half-dollars that could be fully implanted in subjects. His other inventions included an early version of the cardiac pacemaker and implantable "chemitrodes" that could release precise amounts of drugs directly into specific areas of the brain.

In 1952 Delgado coauthored the first peer-reviewed paper describing long-term implantation of electrodes in humans, narrowly beating a report by Robert Heath of Tulane University. Over the next two decades Delgado implanted electrodes in some twenty-five human subjects, most of them schizophrenics and epileptics, at a now defunct mental hospital in Rhode Island. He operated, he says, only on desperately ill patients whose disorders had resisted all previous treatments. Early on, his placement of electrodes in humans was guided by animal experiments, studies of brain-damaged people, and the work of the Canadian neurosurgeon Wilder Penfield; beginning in the 1930s, Penfield stimulated epileptics' brains with electrodes before surgery to determine where he should operate.

Taming a Fighting Bull

Delgado showed that stimulation of the motor cortex could elicit specific physical reactions, such as movement of the limbs. One patient clenched his fist when stimulated, even when he tried to resist. "I guess, doctor, that your electricity is stronger than my will," the patient commented. Another subject, turning his head from side to side in response to stimulation, insisted he was doing so voluntarily, explaining, "I am looking for my slippers."

By stimulating different regions of the limbic system, which regulates emotion, Delgado could also induce fear, rage, lust, hilarity, garrulousness, and other reactions, some of them startling in their intensity. In one experiment, Delgado and two collaborators at Harvard University stimulated the temporal lobe of a twenty-one-year-old epileptic woman while she was calmly playing a guitar; in response, she flew into a rage and smashed her guitar against a wall, narrowly missing a researcher's head.

Perhaps the most medically promising finding was that stimulation of a limbic region called the septum could trigger euphoria, strong enough in some cases to counteract depression and even physical pain. Delgado limited his human research, however, because the therapeutic benefits of implants were unreliable; results varied widely from patient to patient and could be unpredictable even in the same subject. In fact, Delgado recalls turning away more patients than he treated, including a young woman who was sexually promiscuous and prone to violence and had repeatedly been confined in jails and mental hospitals. Although both the woman and her parents begged Delgado to implant electrodes in her, he refused, feeling that electrical stimulation was too primitive for a case involving no discernible neurological disorder.

Delgado did much more extensive research on monkeys and other animals, often focusing on neural regions that elicit and inhibit aggression. In one demonstration, which explored the effects of stimulation on social hierarchy, he implanted a stimoceiver in a macaque bully. He then installed a lever in the cage that, when pressed, pacified the bully by causing the stimoceiver to stimulate the monkey's caudate nucleus, a brain region involved in controlling voluntary movements. A female in the cage soon discovered the lever's power and yanked it whenever the male threatened her. Delgado, who never shied from anthropomorphic interpretations, wrote, "The old dream of an individual overpowering the strength of a dictator by remote control has been fulfilled, at least in our monkey colonies."

Delgado's most famous experiment took place in 1963 at a bull-breeding ranch in Córdoba, Spain. After inserting stimoceivers into the brains of several "fighting bulls," he stood in a bullring with one bull at a time and, by pressing buttons on a hand-held transmitter, controlled each animal's actions. In one instance, captured in a dramatic photograph, Delgado forced a charging bull to skid to a halt only a few feet away from him by stimulating its caudate nucleus. The *New York Times* published a front-page story on the event, calling it "the most spectacular demonstration ever performed of the deliberate modification of animal behavior through external control of the brain." Other articles hailed Delgado's transformation of an aggressive beast into a real-life version of Ferdinand the bull, the gentle hero of a popular children's story.

In terms of scientific significance, Delgado believes his experiment on a female chimpanzee named Paddy deserved more attention. Delgado programmed Paddy's stimoceiver to detect distinctive signals, called spindles, spontaneously emitted by her amygdala. Whenever the stimoceiver detected a spindle, it stimulated the central gray region of Paddy's brain, producing an "aversive reaction" — that is, a painful or unpleasant sensation. After two hours of this negative feedback, Paddy's amygdala produced 50 percent fewer spindles; the frequency dropped by 99 percent within six days. Paddy was not exactly a picture of health: she became "quieter, less attentive and less motivated during behavioral testing," Delgado wrote. He nonetheless speculated that this "automatic learning" technique could be used to quell epileptic seizures, panic attacks, or other disorders characterized by specific brain signals.

Delgado's research was supported not only by civilian agencies but also by military ones such as the Office of Naval Research (but never, Delgado insists, by the Central Intelligence Agency, as some conspiracy theorists have charged). Delgado, who calls himself a pacifist, says that his Pentagon sponsors viewed his work as basic research and never steered him toward military applications. He has always dismissed speculation that implants could create cyborg soldiers who kill on command, like the brainwashed assassin in the novel and film versions of *The Manchurian Candidate.* (The assassin was controlled by psychological methods in the original 1962 film and by a brain chip in the 2004 remake.) Brain stimulation may "increase or decrease aggressive behavior," he asserts, but it cannot "direct aggressive behavior to any specific target."

Envisioning a "Psychocivilized Society"

In 1969 Delgado described brain-stimulation research and discussed its implications in *Physical Control of the Mind: Toward a Psychocivilized Society,* which was illustrated with photographs of monkeys, cats, a bull, and two young women whose turbans concealed stimoceivers. (Female patients "have shown their feminine adaptability to circumstance," Delgado remarked, "by wearing attractive hats or wigs to conceal their electrical headgear.") Spelling out the limitations of brain stimulation, Delgado downplayed "Orwellian possibilities" in which evil scientists enslave people by implanting electrodes in their brains.

Yet some of his rhetoric had an alarmingly evangelical tone. Neurotechnology, he declared, was on the verge of "conquering the mind" and creating "a less cruel, happier, and better man." In a review in *Scientific American,* the late physicist Philip Morrison called *Physical Control* "a thoughtful, up-to-date account" of electrical stimulation experiments but added that its implications were "somehow ominous."

In 1970 Delgado's field was engulfed in a scandal triggered by Frank Ervin and Vernon Mark, two researchers at Harvard Medical School with whom Delgado briefly collaborated. (One of Ervin's students was Michael Crichton, who wrote *The Terminal Man.* The bestseller, about a bionic experiment gone awry, was inspired by the research of Ervin, Mark, and Delgado.) In their book *Violence and the Brain,* Ervin and Mark suggested that brain stimulation or psychosurgery might quell the violent tendencies of blacks rioting in inner cities. In 1972 Heath, the Tulane psychiatrist, raised more questions about brain-implant research when he reported that he had tried to change the sexual orientation of a male homosexual by stimulating his septal region while he had intercourse with a female prostitute.

The fiercest opponent of brain implants was the psychiatrist Peter Breggin (who in recent decades has focused on the dangers of psychiatric drugs). In testimony entered into the *Congressional Record* in 1972, Breggin lumped Delgado, Ervin, Mark, and Heath together with advocates of lobotomies and accused them of trying to create "a society in which everyone who deviates from the norm" will be "surgically mutilated." Quoting liberally from *Physical Control,* Breggin singled out Delgado as "the great apologist for technologic totalitarianism." In his 1973 book *Brain Control,* Elliot Valenstein, a neurophysiologist at the University of Michigan at Ann Arbor, presented a detailed scientific critique of brain-implant research by Delgado and others, contending that the results of stimulation were much less precise and therapeutically beneficial than proponents often suggested. (Delgado notes that in his own writings he made many of the same points as Valenstein.)

Meanwhile strangers started accusing Delgado of having secretly implanted stimoceivers in their brains. One woman who made this claim sued Delgado and Yale University for one million dollars, although he had never met her. In the midst of this brouhaha, Villar Palasí, the Spanish minister of health, asked Delgado to help organize a new medical school at the Autonomous University in Madrid, and he accepted, moving with his wife and two children to Spain in 1974. He insists that he was not fleeing the disputes surrounding his research; the minister's offer was just too good to refuse. "I said, 'Could I have the facilities I have at Yale?' And he said, 'Oh, no, much better!' "

In Spain, Delgado shifted his focus to noninvasive methods of affecting the brain, which he hoped would be more medically acceptable than implants. Anticipating modern techniques such as transcranial magnetic stimulation, he invented a halolike device and a helmet that could deliver electromagnetic pulses to specific neural regions. Testing the gadgets on both animals and human volunteers — including himself and his daughter, Linda — Delgado discovered that he could induce drowsiness, alertness, and other states; he also had some success treating tremors in Parkinson's patients.

Delgado still could not entirely escape controversy. In the mid-1980s an article in the magazine *Omni* and documentaries by the BBC and CNN cited Delgado's work as circumstantial evidence that the United States and the Soviet Union might have secretly developed methods for remotely modifying people's thoughts. Noting that the power and precision of electromagnetic pulses decline rapidly with distance, Delgado dismisses these mind-control claims as "science fiction."

Except for these flashes of publicity, however, Delgado's work no longer received the attention it once had. Although he continued publishing articles — especially on the effects of electromagnetic radiation on cognition, behavior, and embryonic growth — many appeared only in Spanish journals. Moreover, brain-stimulation studies in the United States bogged down in ethical controversies, grants dried up, and researchers drifted to other fields, notably psychopharmacology, which seemed to be a much safer, more effective way to treat brain disorders than brain stimulation or surgery. Only in the past decade has brain-implant research revived, spurred by advances in computation, electrodes, microelectronics, and brain-scanning technologies — and by a growing recognition of the limits of drugs for treating mental illness.

Delgado, who stopped doing research in the early 1990s but still follows the field of brain stimulation, believes modern investigators fail to cite his studies not because he was so controversial but simply out of ignorance; after all, most modern databases do not include publications from his heyday in the 1950s and 1960s. He is thrilled by the resurgence of research on brain stimulation, because he still believes in its potential to liberate us from psychiatric diseases and our innate aggression. "In the near future," he says, "I think we will be able to help many human beings, especially with the noninvasive methods."

Delgado's successors have faced some of the same questions that he did about possible abuses of neurotechnology. Some pundits have expressed concern that brain chips could allow a "controlling organization" to "hack into the wetware between our ears," as *New York Times* columnist William Safire put it. An editorial in *Nature* recently expressed concern that officials in the Defense Advanced Research Projects Agency, a major funder of brain-implant research, have openly considered implanting brain chips in soldiers to boost their cognitive capacities. Meanwhile some technoenthusiasts, such as the British computer scientist Kevin Warwick, contend that the risks of brain chips are far outweighed by the potential benefits, which will include instantly "downloading" new languages or other skills, controlling computers and other devices with our thoughts, and communicating telepathically with one another.

Delgado predicts that neurotechnologies may never advance as far as many people fear or hope. The applications envisioned by Warwick and others, Delgado points out, require knowing how complex information is encoded in the brain, a goal that neuroscientists are far from achieving. Moreover, learning quantum mechanics or a new language involves "slowly changing connections which are already there," Delgado says. "I don't think you can do that suddenly." Brain stimulation, he adds, can only modify skills and capacities that we already possess.

But Delgado looks askance at the suggestion of the White House Council on Bioethics and others that some scientific goals — particularly those that involve altering human nature — should not even be pursued. To be sure, he says, technology "has two sides, for good and for bad," and we should do what we can to "avoid the adverse consequences." We should try to prevent potentially destructive technologies from being abused by authoritarian governments to gain more power or by terrorists to wreak destruction. But human nature, Delgado asserts, echoing one of the themes of *Physical Control,* is not static but "dynamic," constantly changing as a result of our compulsive self-exploration. "Can you avoid knowledge?" Delgado asks. "You cannot! Can you avoid technology? You cannot! Things are going to go ahead in spite of ethics, in spite of your personal beliefs, in spite of everything."

GORDON KANE

The Mysteries of Mass

FROM *Scientific American*

MOST PEOPLE THINK they know what mass is, but they understand only part of the story. For instance, an elephant is clearly bulkier and weighs more than an ant. Even in the absence of gravity, the elephant would have greater mass — it would be harder to push and set in motion. Obviously the elephant is more massive because it is made of many more atoms than the ant is, but what determines the masses of the individual atoms? What about the elementary particles that make up the atoms — what determines their masses? Indeed, why do they even have mass?

We see that the problem of mass has two independent aspects. First, we need to learn how mass arises at all. It turns out mass results from at least three different mechanisms, which I will describe below. A key player in physicists' tentative theories about mass is a new kind of field that permeates all of reality, called the Higgs field. Elementary particle masses are thought to come about from interaction with the Higgs field. If the Higgs field exists, theory demands that it have an associated particle, the Higgs boson. Using particle accelerators, scientists are now hunting for the Higgs.

The second aspect is that scientists want to know why different species of elementary particles have their specific quantities of mass. Their intrinsic masses span at least eleven orders of magnitude, but we do not yet know why that should be so. For comparison, an elephant and the smallest of ants differ by about eleven orders of magnitude of mass.

What Is *Mass?*

Isaac Newton presented the earliest scientific definition of mass in 1687 in his landmark *Principia:* "The quantity of matter is the measure of the same, arising from its density and bulk conjointly." That very basic definition was good enough for Newton and other scientists for more than two hundred years. They understood that science should proceed first by describing how things work and later by understanding why. In recent years, however, the *why* of mass has become a research topic in physics. Understanding the meaning and origins of mass will complete and extend the Standard Model of particle physics, the well-established theory that describes the known elementary particles and their interactions. It will also resolve mysteries such as dark matter, which makes up about 25 percent of the universe.

The foundation of our modern understanding of mass is far more intricate than Newton's definition and is based on the Standard Model. At the heart of the Standard Model is a mathematical function called a Lagrangian, which represents how the various particles interact. From that function, by following rules known as relativistic quantum theory, physicists can calculate the behavior of the elementary particles, including how they come together to form compound particles, such as protons. For both the elementary and the compound particles, we can then calculate how they will respond to forces, and for a force F, we can write Newton's equation $F = ma$, which relates the force, the mass, and the resulting acceleration. The Lagrangian tells us what to use for m here, and that is what is meant by the mass of the particle.

But mass, as we ordinarily understand it, shows up in more than just $F = ma$. For example, Einstein's special relativity theory predicts that massless particles in a vacuum travel at the speed of light and that particles with mass travel more slowly, in a way that can be calculated if we know their mass. The laws of gravity predict that gravity acts on mass and energy as well, in a precise manner. The quantity m deduced from the Lagrangian for each particle behaves correctly in all those ways, just as we expect for a given mass.

Fundamental particles have an intrinsic mass known as their rest mass (those with zero rest mass are called massless). For a compound particle, the constituents' rest mass and also their kinetic energy of motion and potential energy of interactions contribute to the particle's total mass. Energy and mass are related, as described by Einstein's famous equation, $E = mc^2$ (energy equals mass times the speed of light squared).

An example of energy contributing to mass occurs in the most familiar kind of matter in the universe — the protons and neutrons that make up atomic nuclei in stars, planets, people, and all that we see. These particles amount to 4 to 5 percent of the mass-energy of the universe. The Standard Model tells us that protons and neutrons are composed of elementary particles called quarks that are bound together by massless particles called gluons. Although the constituents are whirling around inside each proton, from outside we see a proton as a coherent object with an intrinsic mass, which is given by adding up the masses and energies of its constituents.

The Standard Model lets us calculate that nearly all the mass of protons and neutrons is from the kinetic energy of their constituent quarks and gluons (the remainder is from the quarks' rest mass). Thus, about 4 to 5 percent of the entire universe — almost all the familiar matter around us — comes from the energy of motion of quarks and gluons in protons and neutrons.

The Higgs Mechanism

Unlike protons and neutrons, truly elementary particles — such as quarks and electrons — are not made up of smaller pieces. The explanation of how they acquire their rest masses gets to the very heart of the problem of the origin of mass. As I noted above, the account proposed by contemporary theoretical physics is that fundamental particle masses arise from interactions with the Higgs field. But why is the Higgs field present throughout the universe? Why isn't its strength essentially zero on cosmic scales, like the electromagnetic field? What is the Higgs field?

The Higgs field is a quantum field. That may sound mysterious, but the fact is that all elementary particles arise as quanta of a corresponding quantum field. The electromagnetic field is also a quantum field (its corresponding elementary particle is the photon). So in this respect, the Higgs field is no more enigmatic than electrons and light. The Higgs field does, however, differ from all other quantum fields in three crucial ways.

The first difference is somewhat technical. All fields have a property called spin, an intrinsic quantity of angular momentum that is carried by each of their particles. Particles such as electrons have spin ½, and most particles associated with a force, such as the photon, have spin 1. The Higgs boson (the particle of the Higgs field) has spin 0. Having 0 spin enables the Higgs field to appear in the Lagrangian in different ways from the other particles, which in turn allows — and leads to — its other two distinguishing features.

The second unique property of the Higgs field explains how and why it has nonzero strength throughout the universe. Any system, including a universe, will tumble into its lowest energy state, like a ball bouncing down to the bottom of a valley. For the familiar fields, such as the electromagnetic fields that give us radio broadcasts, the lowest energy state is the one in which the fields have zero value (that is, the fields vanish) — if any nonzero field is introduced, the energy stored in the fields increases the net energy of the system. But for the Higgs field, the energy of the universe is lower if the field is not zero but instead has a constant nonzero value. In terms of the valley metaphor, for ordinary fields the valley floor is at the location of zero field; for the Higgs, the valley has a hillock at its center (at zero field), and the lowest point of the valley forms a circle around the hillock. The universe, like a ball, comes to rest somewhere on this circular trench, which corresponds to a nonzero value of the field. That is, in its natural, lowest energy state, the universe is permeated throughout by a *nonzero* Higgs field.

The final distinguishing characteristic of the Higgs field is the form of its interactions with the other particles. Particles that interact with the Higgs field behave as if they have mass, proportional to the strength of the field times the strength of the interaction. The masses arise from the terms in the Lagrangian that have the particles interacting with the Higgs field.

Our understanding of all this is not yet complete, however, and we are not sure how many kinds of Higgs fields there are. Although the Standard Model requires only one Higgs field to generate all the elementary particle masses, physicists know that the Standard Model must be superseded by a more complete theory. Leading contenders are extensions of the Standard Model known as Supersymmetric Standard Models (SSMs). In these models, each Standard Model particle has a so-called superpartner (as yet undetected) with closely related properties. With the Supersymmetric

Standard Model, at least two different kinds of Higgs fields are needed. Interactions with those two fields give mass to the Standard Model particles. They also give some (but not all) mass to the superpartners. The two Higgs fields give rise to five species of Higgs boson: three that are electrically neutral and two that are charged. The masses of particles called neutrinos, which are tiny compared with other particle masses, could arise rather indirectly from these interactions or from yet a third kind of Higgs field.

Theorists have several reasons for expecting the SSM picture of the Higgs interaction to be correct. First, without the Higgs mechanism, the *W* and *Z* bosons that mediate the weak force would be massless, just like the photon (which they are related to), and the weak interaction would be as strong as the electromagnetic one. Theory holds that the Higgs mechanism confers mass to the *W* and *Z* in a very special manner. Predictions of that approach (such as the ratio of the *W* and *Z* masses) have been confirmed experimentally.

Second, essentially all other aspects of the Standard Model have been well tested, and with such a detailed, interlocking theory it is difficult to change one part (such as the Higgs) without affecting the rest. For example, the analysis of precision measurements of *W* and *Z* boson properties led to the accurate prediction of the top quark mass before the top quark had been directly produced. Changing the Higgs mechanism would spoil that and other successful predictions.

Third, the Standard Model Higgs mechanism works very well for giving mass to *all* the Standard Model particles, *W* and *Z* bosons, as well as quarks and leptons; the alternative proposals usually do not. Next, unlike the other theories, the SSM provides a framework to unify our understanding of the forces of nature. Finally, the SSM can explain why the energy "valley" for the universe has the shape needed by the Higgs mechanism. In the basic Standard Model the shape of the valley has to be put in as a postulate, but in the SSM that shape can be derived mathematically.

Testing the Theory

Naturally, physicists want to carry out direct tests of the idea that mass arises from the interactions with the different Higgs fields. We can test three key features. First, we can look for the signature particles called Higgs bosons. These quanta must exist or else the explanation is not right. Physicists are currently looking for Higgs bosons at the Tevatron Collider at Fermi National Accelerator Laboratory in Batavia, Illinois.

Second, once they are detected, we can observe how Higgs bosons interact with other particles. The very same terms in the Lagrangian that determine the masses of the particles also fix the properties of such interactions. So we can conduct experiments to test quantitatively the presence of interaction terms of that type. The strength of the interaction and the amount of particle mass are uniquely connected.

Third, different sets of Higgs fields, as occur in the Standard Model or in the various SSMs, imply different sets of Higgs bosons with various properties, so tests can distinguish these alternatives, too. All that we need to carry out the tests is appropriate particle colliders — ones that have sufficient energy to produce the different Higgs bosons, sufficient intensity to make enough of them, and very good detectors to analyze what is produced.

A practical problem with performing such tests is that we do not yet understand the theories well enough to calculate what masses the Higgs bosons themselves should have, which makes searching for them more difficult because one must examine a range of masses. A combination of theoretical reasoning and data from experiments guides us about roughly what masses to expect.

The Large Electron-Positron Collider (LEP) at CERN, the European laboratory for particle physics near Geneva, operated over a mass range that had a significant chance of including a Higgs boson. It did not find one — although there was tantalizing evidence for one just at the limits of the collider's energy and intensity — before it was shut down in 2000 to make room for construction of a newer facility, CERN's Large Hadron Collider (LHC). The Higgs must therefore be heavier than about 120 proton masses. Nevertheless, LEP did produce indirect evidence that a Higgs boson exists: experimenters at LEP made a number of precise measurements, which can be combined with similar measurements from the Tevatron and the collider at the Stanford Linear Accelerator Center. The entire set of data agrees well with theory only if certain interactions of particles with the lightest Higgs boson are included and only if the lightest Higgs boson is not heavier than about 200 proton masses. That provides researchers with an upper limit for the mass of the Higgs boson, which helps focus the search.

For the next few years, the only collider able to produce direct evidence for Higgs bosons will be the Tevatron. Its energy is sufficient to discover a Higgs boson in the range of masses implied by the indirect LEP evidence *if* it can consistently achieve the beam intensity it was expected to have, which so far has not been possible. In 2007 the LHC, which is seven times more energetic and is designed to have far more intensity than the Tevatron, is scheduled to begin taking data. It will be a factory for Higgs bosons (meaning it will produce many of the particles a day). Assuming the LHC functions as planned, gathering the relevant data and learning how to interpret it should take one to two years. Carrying out the complete tests that show in detail that the interactions with Higgs fields are providing the mass will require a new electron-positron collider in addition to the LHC (which collides protons) and the Tevatron (which collides protons and antiprotons).

Dark Matter

What is discovered about Higgs bosons will not only test whether the Higgs mechanism is indeed providing mass, it will also point the way to how the Standard Model can be extended to solve problems such as the origin of dark matter.

With regard to dark matter, a key particle of the SSM is the lightest superpartner (LSP). Among the superpartners of the known Standard Model particles predicted by the SSM, the LSP is the one with the lowest mass. Most superpartners decay promptly to lower-mass superpartners, a chain of decays that ends with the LSP, which is stable because it has no lighter particle that it can decay into. (When a superpartner decays, at least one of the decay products should be another superpartner; it should not decay entirely into Standard Model particles.) Superpartner particles would have been created early in the big bang but then promptly decayed into LSPs. The LSP is the leading candidate particle for dark matter.

The Higgs bosons may also directly affect the amount of dark matter in the universe. We know that the amount of LSPs today should be less than the amount shortly after the big bang, because some would have collided and annihilated into quarks and leptons and photons, and the annihilation rate may be dominated by LSPs interacting with Higgs bosons.

As mentioned earlier, the two basic SSM Higgs fields give mass to the Standard Model particles and *some* mass to the superpartners, such as the LSP. The superpartners acquire more mass via additional interactions, which may be with still further Higgs fields or with fields similar to the Higgs. We have theoretical models of how these processes can happen, but until we have data on the superpartners themselves we will not know how they work in detail. Such data are expected from the LHC or perhaps even from the Tevatron.

Neutrino masses may also arise from interactions with additional Higgs or Higgs-like fields, in a very interesting way. Neutrinos were originally assumed to be massless, but since 1979 theorists have predicted that they have small masses, and over the past decade several impressive experiments have confirmed the predictions. The neutrino masses are less than a millionth the size of the next smallest mass, the electron mass. Because neutrinos are electrically neutral, the theoretical description of their masses is more subtle than for charged particles. Several processes contribute to the mass of each neutrino species, and for technical reasons the actual mass value emerges from solving an equation rather than just adding the terms.

Thus, we have understood the three ways that mass arises: the main form of mass we are familiar with — that of protons and neutrons and therefore of atoms — comes from the motion of quarks bound into protons and neutrons. The proton mass would be about what it is even without the Higgs field. The masses of the quarks themselves, however, and also the mass of the electron, are entirely caused by the Higgs field. Those masses would vanish without the Higgs. Last, but certainly not least, most of the amount of superpartner masses, and therefore the mass of the dark-matter particle (if it is indeed the lightest superpartner), comes from additional interactions beyond the basic Higgs one.

Finally, we consider an issue known as the family problem. Over the past half a century physicists have shown that the world we see, from people to flowers to stars, is constructed from just six particles: three matter particles (up quarks, down quarks, and electrons), two force quanta (photons and gluons), and Higgs bosons

— a remarkable and surprisingly simple description. Yet there are four more quarks, two more particles similar to the electron, and three neutrinos. All are very short-lived or barely interact with the other six particles. They can be classified into three families: up, down, electron neutrino, electron; charm, strange, muon neutrino, muon; and top, bottom, tau neutrino, tau. The particles in each family have interactions identical to those of the particles in other families. They differ only in that those in the second family are heavier than those in the first, and those in the third family are heavier still. Because these masses arise from interactions with the Higgs field, the particles must have different interactions with the Higgs field.

Hence, the family problem has two parts. Why are there three families when it seems only one is needed to describe the world we see? Why do the families differ in mass and have the masses they do? Perhaps it is not obvious why physicists are astonished that nature contains three almost identical families even if one would do. It is because we want to fully understand the laws of nature and the basic particles and forces. We expect that every aspect of the basic laws is a necessary one. The goal is to have a theory in which all the particles and their mass ratios emerge inevitably, without making ad hoc assumptions about the values of the masses and without adjusting parameters. If having three families is essential, then it is a clue whose significance is currently not understood.

Tying It All Together

The standard model and the SSM can accommodate the observed family structure, but they cannot explain it. This is a strong statement. It is not that the SSM has not *yet* explained the family structure but that it *cannot.* For me, the most exciting aspect of string theory is not only that it may provide us with a quantum theory of all the forces but also that it may tell us what the elementary particles are and why there are three families. String theory seems able to address the question of why the interactions with the Higgs field differ among the families. In string theory, repeated families can occur, and they are not identical. Their differences are described by properties that do not affect the strong, weak, electromagnetic, or gravitational forces but that do affect the interactions with Higgs fields, which fits with our having three families with different masses. Although string theorists have not yet fully solved the problem of having three families, the theory seems to have the right structure to provide a solution. String theory allows many different family structures, and so far no one knows why nature picks the one we observe rather than some other. Data on the quark and lepton masses and on their superpartner masses may provide major clues to teach us about string theory.

One can now understand why it took so long historically to begin to understand mass. Without the Standard Model of particle physics and the development of quantum field theory to describe particles and their interactions, physicists could not even formulate the right questions. Whereas the origins and values of mass are not yet fully understood, it is likely that the framework needed to understand them is in place. Mass could not have been comprehended before theories such as the Standard Model and its supersymmetric extension and string theory existed. Whether they indeed provide the complete answer is not yet clear, but mass is now a routine research topic in particle physics.

KEVIN KRAJICK

Future Shocks

FROM *Smithsonian*

BRIAN ATWATER paddled a battered aluminum canoe up the Copalis River, pushed along by a rising Pacific tide. At this point, a 130-mile drive from Seattle, the hundred-foot-wide river wound through wide salt marshes fringed with conifers growing on high ground. The scene, softened by gray winter light and drizzle, was so quiet one could hear the whisper of surf a mile away. But then Atwater rounded a bend, and a vision of sudden, violent destruction appeared before him: stranded in the middle of a marsh were dozens of towering western red cedars, weathered like old bones, their gnarly, hollow trunks wide enough to crawl into. "The ghost forest," Atwater said, pulling his paddle from the water. "Earthquake victims."

Atwater beached the canoe and got out to walk among the spectral giants, relics of the last great Pacific Northwest earthquake. The quake generated a vast tsunami that inundated parts of the West Coast and surged across the Pacific, flooding villages some 4,500 miles away in Japan. It was as powerful as the one that killed more than 220,000 people in the Indian Ocean in December 2004. The cedars died after saltwater rushed in, poisoning their roots but leaving their trunks standing. This quake is not noted in any written North American record, but it is clearly written in the earth. The ghost forest stands as perhaps the most conspicuous and haunting warning that it has happened here before — and will surely happen here again.

"When I started out, a lot of these dangers were not all that clear," says Atwater, a geologist for the U.S. Geological Survey

(USGS) who specializes in the science of paleoseismology, or the study of earthquakes past. "If you look at what we know now, it beats you over the head."

In one of the more remarkable feats of modern geoscience, researchers have pinpointed the date, hour, and size of the cataclysm that killed these cedars. In Japan, officials recorded an "orphan" tsunami — unconnected with any felt earthquake — with waves up to ten feet high along six hundred miles of the Honshu coast at midnight, January 27, 1700. Several years ago, Japanese researchers, by estimating the tsunami's speed, path, and other properties, concluded that it was triggered by a magnitude 9 earthquake that warped the sea floor off the Washington coast at 9 P.M. Pacific Standard Time on January 26, 1700. To confirm it, U.S. researchers found a few old trees of known age that had survived the earthquake and compared their tree rings with the rings of the ghost-forest cedars. The trees had indeed died just before the growing season of 1700.

In the Pacific Northwest, where written records start in the late 1700s, paleoseismologists have spotted many other signs of past disasters, from sands washed far inshore to undersea landslides. In addition to the risk from offshore earthquakes, recent studies show that Seattle and the greater Puget Sound area, with its four million people, is itself underlain by a network of faults in the earth's surface. They also have ruptured catastrophically in the not very distant past. Considering all the geologic evidence, scientists now say a major earthquake strikes the Pacific Northwest every few hundred years — give or take a few hundred years. That means the next one could strike tomorrow.

The study of the past has taken on paramount importance because scientists still cannot predict earthquakes, though not for lack of effort. One important quake-forecasting experiment has taken place since 1985 in tiny Parkfield, California, the self-proclaimed "earthquake capital of the world." The town sits atop a highly active section of the San Andreas Fault, the dangerous crack that cuts the state south to north for 800 miles. Underlying geological forces cause quakes to occur in the same places repeatedly. Until recently, much of modern earthquake theory was based on the idea that intervals between these events were nicely regular. Through most of the twentieth century, Parkfield, for example, had one every twenty-two years or so. But experience now shows that quakes are maddeningly unpredictable. Scientists forecast that a quake would hit Parkfield in 1988, give or take five years. They installed networks of strainmeters, creepmeters, seismometers, and other instruments around the town. Their goal was to capture precursors to the expected quake, such as a pattern of subtle tremors, which they could later use to predict when another quake is imminent. The earthquake did come along — in September 2004, with one-twentieth the expected power — and with no warning whatsoever. Looking at all their measurements, scientists still have found no reliable signs that an earthquake is about to strike.

Still, by gathering ever more information about the past, paleoseismologists are becoming adept at mapping danger zones and spreading the warning, even if they can't say when the next one is due. The information, though imprecise, is useful to engineers, city planners, and others, who can strengthen building codes and educate the public about how to survive a major quake whenever it comes. Art Frankel, a chief architect of the USGS national seismic hazard mapping project, says such geological "hazard maps" are like charts of the most dangerous traffic intersections; they can't predict when the next car accident will happen, but they do tell you to watch out.

Because of these studies of past earthquakes, the world is looking ever more inhospitable. Paleoseismology is turning up portentous signs of past upheavals in the American Midwest, eastern Canada, Australia, and Germany. "We're discovering some new hazard every few months," says Brian Sherrod, a USGS geologist investigating the Seattle faults. The Pacific Northwest may not be the only place harboring such nasty surprises, but it is where the geological signs are most dramatic, the science is moving fast, and a future earthquake would be among the most catastrophic.

The earth's crust consists of interlocking tectonic plates floating on the hot, pliable interior of the planet, which drift and collide with one another. The Pacific Northwest coast is such a dangerous place because it rests on a continental plate that meets, some thirty to ninety miles offshore, a sea-floor plate. The boundary between the two plates, stretching seven hundred miles from British Columbia to Northern California, is called the Cascadia subduction zone.

Subduction is the process by which an ocean plate nudges under a continental plate, usually by a few inches a year. Grinding between such plates can bring small temblors, but often the parts lock against each other like sticky watch gears, causing the still advancing seafloor to compress like a spring and the overlying coastline to warp upward. When the pent-up pressure finally pops, the sea floor lunges landward and the coast lunges seaward, with seaside real estate collapsing. The shifting plates displace seawater in all directions, creating a tsunami that travels up to five hundred miles an hour. These subduction-zone quakes are the world's largest, dwarfing those that take place in the land's crust. December's subduction quake in Indonesia, a magnitude 9, was about thirty times more powerful than the 1906 San Francisco event, which took place in the continental crust near the city. Other major subduction-zone quakes off Alaska in 1946 and 1964 sent tsunamis all the way to Hawaii and Northern California, killing scores of people.

Downriver of the ghost forest, with heavy rain threatening the tidal estuary of the Copalis River, Atwater stepped from the canoe to stand crotch-deep in cold water and mud. He wore hiking boots and chest waders, having learned long ago that tidal mud can suck hip waders right off you. Wielding an entrenching tool, a military folding shovel, he chopped at the riverbank to view the sedimentary layers, which can yield a great deal of information about past quakes. Every time a sea-floor earthquake occurs here, forests and marshes suddenly drop and are reburied by later sediments washed in by tides and river drainage. A geologist can dig a hole in search of such buried evidence — or find a riverbank where erosion has done most of the work for him, which was what Atwater had here. His tool kit also included a hunting knife and a *nejiri gama,* a trowel-size Japanese gardening tool shaped like a hoe.

Atwater kneeled in the shallows and scraped riverbank mud down onto his thighs, then smoothed the bank with the nejiri gama. Below the two and half feet of brownish tidal muck lay a half-inch band of gray sand, which was neatly draped over black peat. The peat was laced with tree roots, even though the nearest visible tree was far across the marsh. "Hoo, that's nice, that's fresh!" Atwater shouted. "Old dependable!" These trees grow only above the tide line and were now below it. Something, he said, had dropped this ecosystem several feet all at once; all signs point to a sea-floor quake. Radiocarbon dating has shown that the plants died about three hundred years ago. The overlying sand sheet was the clincher: only a tsunami could have laid it down.

Atwater, fifty-three, has been combing the region since 1986 for evidence of past earthquakes, and his work at a dozen estuaries — in addition to other scientists' findings — has revealed not only the great 1700 earthquake and tsunami but also a dozen other major quakes over the past seven thousand years. Recent sea-floor studies off the Pacific Northwest coast tell the same story. Overall, big subduction-zone quakes strike on average every five hundred to six hundred years. But the intervals between them range from two hundred to a thousand years. "*If* we can predict that we're in a short interval, we've essentially used up our time. But we *can't* predict," says Chris Goldfinger, a marine geologist at Oregon State University. Recent studies using satellite-controlled global-positioning systems and other new technology confirm that the region's tectonic plates are converging and locked together. In some places, the Washington and Oregon coastlines are rising by 4 millimeters a year. As Atwater points out, "That doesn't sound like much until you multiply it by, say, one thousand years, and you get ten feet." And if the land has risen that far, it could drop that far when a quake comes, just like the layer of peat Atwater uncovered in the tidal estuary. "The bulge will collapse during the next earthquake, and there will be new ghost forests," he says.

We paddled farther up the Copalis to the mouth of a small creek, where Atwater located the 1700 tsunami sand sheet's continuation in the riverbank. With his nejiri gama, he dug out clumps of perfectly preserved ancient spruce needles, apparently cast up by the great waves. Nearby he uncovered a shard of fire-cracked rock — evidence of a cook fire. "That's spooky," he says. "It makes you wonder what happened to these people." Paleoseismology has shed new light on legends of aboriginal coastal peoples such as the Yurok and the Quileute. Many stories describe times when the earth shook and the ocean crashed in, wiping out villages, stranding canoes in trees, and killing everyone but the fastest or luckiest. Storytellers often explained these events as the result of a battle between a great whale and a thunderbird. "Well before settlers came here, native peoples dealt with earthquakes," says James Rasmussen, a councilman for the Duwamish people in Seattle. Archaeologists have now identified many sites that contain pottery and other artifacts that were submerged by rising waters. Apparently, native people over the years moved closer to the shore or fled it as thunderbird and whale fought it out.

Today, of course, we're not so light on our feet. A recent study estimates that ten million people on the West Coast would be affected by a Cascadia subduction–zone quake. Three hundred years of tectonic pressure have now built up. The shaking from such a quake, lasting two to four minutes, would damage two hundred highway bridges, put Pacific ports out of business for months, and generate low-frequency shock waves possibly capable of toppling tall buildings and long bridges in Seattle and Portland, Oregon. A tsunami of thirty feet or more would reach parts of the Pacific Coast in little over half an hour. Of special concern to Washington State officials are places like the coastal resort town of Ocean Shores, on a long sand spit with a narrow access road that serves fifty thousand visitors on a summer day. Here the highest ground — twenty-six feet above sea level — would hold only "about one hundred people who are very good friends," says Tim Walsh, the state geological-hazards program manager. He suggests that the town consider "vertical evacuation" — building multistory schools or other public structures in which people on the top floors could escape a tsunami, assuming the buildings themselves could withstand the impact. To flee a tsunami, people need warnings, and the U.S. government has set out Pacific Ocean monitors to pick up signals from known danger spots, not only in the Pacific Northwest but in Japan, Russia, Chile, and Alaska as well. This system is designed to transmit warnings to countries across the basin within minutes. Similar networks are planned for the Atlantic and Indian oceans.

In Washington State, officials are trying to educate a public that has regarded the threat casually — but that now, with the Indian Ocean tsunami as an object lesson, may pay a lot more attention. A few weeks before the disaster, Atwater and Walsh drove to Port Townsend, a Victorian-era seaport on the Strait of Juan de Fuca, about midway between Seattle and the open ocean, where they ran a tsunami workshop that was attended by only a handful of emergency officials and a few dozen residents. Walsh pointed out that a tsunami might take a couple of hours to reach Port Townsend, which has nearby cliffs for retreat. The town is dotted with blue-and white tsunami warning signs. Unfortunately, they are a popular souvenir. "Just please stop stealing the signs," Walsh chided the audience as he handed out free paper replicas of the signs.

"A lot of people think of tsunamis as some kind of cool adventure," Walsh said after the meeting. He remembered that following a big 1994 sea-floor quake off Russia's Kuril Islands, surfers in Hawaii headed for the beaches. A film crew actually set up at the surf line on the Washington coast, hoping to catch a giant wave that, fortunately for them, never came. Walsh said, "I think they won't be doing that next time."

Brian Sherrod, a geologist with the USGS in Seattle, has rush-hour traffic to thank for one discovery. Recently he led some visitors under Interstate 5, a ten-lane raised artery traversing the city's downtown, as thousands of northbound cars and trucks thundered overhead. He pointed to the ground beneath one of the massive concrete supports, where the ruptures of an earthquake fault in prehistoric times had tortured the usually flat sediment layers into broken waves, then smashed and bent them backward so that lower ones were shoved over the upper — as if someone had taken a layer cake and slammed a door on it. This is one of many scary signs from Seattle's past, though one of the few visible to the naked eye. "I spotted this when I was stopped in Friday-afternoon traffic," said Sherrod, pointing to the southbound lanes, fifty feet away at eye level. "I was singing real loud to the radio. Then I stopped singing and yelled, 'Holy shit!'"

Earthquakes have long been a fact of life in Seattle. Each year inland Washington gets a dozen or so quakes big enough to feel, and since 1872, about two dozen have caused damage. Most cluster under the Puget Sound lowland, the heavily developed run of bays, straits, islands, and peninsulas running through Seattle south to Olympia. Larger-than-usual quakes in 1949 and 1965 killed fourteen people. In the past few decades, building codes have been upgraded and a network of seismometers has been installed across Washington and Oregon. Those instruments showed that most of the smaller quakes are shallow readjustments of the earth's crust — rarely a big deal. The more sizable events, like the quakes in

1949 and 1965, typically emanate from depths of thirty miles or more. Fortunately, this is far enough down that a lot of energy bleeds from the seismic shock waves before they reach the surface. The most recent big deep one was the February 28, 2001, Nisqually quake — magnitude 6.8, as measured at its thirty-two-mile-deep point of origin. It damaged older masonry buildings in Seattle's picturesque Pioneer Square shopping district, where unreinforced bricks flattened cars; at the vast nearby cargo harbor, pavement split and sand volcanoes boiled up. Though damage was some $2 billion to $4 billion statewide, many businesses were able to reopen within hours.

One of the first hints that monstrous quakes take place near Seattle's surface, where they can do catastrophic damage, came when companies were hunting for oil under Puget Sound in the 1960s, and geophysicists spotted apparent faults in the sound's floor. Into the 1990s, these were presumed to be inactive relic faults; then scientists looked more closely. At Restoration Point, on populous Bainbridge Island, across Puget Sound from downtown Seattle, one USGS scientist recognized evidence of what geologists call a marine terrace. This is a stairstep structure made up of a wave-cut sea cliff topped by a flat, dry area that runs as much as several hundred feet inland to a similar but higher cliff. Restoration Point's sharp, uneroded edges and the ancient marine fossils found on the flat step suggested that the whole block had risen more than twenty feet from the water all at once. Several miles north of the point lies a former tideland that apparently had dropped at the same time. These paired formations are the signature of what's known as a reverse fault, in which the earth's crust gets shoved up violently on one side and down on the other. This one is now called the Seattle fault zone. It runs west to east for at least forty miles, under Puget Sound, downtown Seattle (cutting it in half) and its suburbs, and nearby lakes.

Along the Seattle fault on the east side of the city, Gordon Jacoby, a Columbia University tree-ring specialist, has identified another ghost forest — under sixty feet of water in Lake Washington. The trees did not sink; they rode off a nearby hill on a gigantic quake-induced landslide in the year 900 AD, apparently at the same time that Restoration Point rose. Yet more evidence of that devastating event surfaced a decade ago several miles north of the Seattle fault. The city was digging a sewer, and Atwater spotted in one of the excavations an inland tsunami deposit — the first of many tied to that quake. The tsunami came when the fault thrust up under Puget Sound, sending out waves that smashed what is now the booming metropolitan waterfront.

Geologists have spotted at least five other fault zones in the region, from the Canadian border south to Olympia. The faults bear signs of half a dozen ruptures over the past twenty-five hundred years, and one fault, the Utsalady, just north of Seattle, might have ruptured as recently as the early 1800s. The evidence amassed so far suggests an average repeat time for a major shallow continental earthquake from centuries to millennia.

The USGS has mounted a campaign to map the faults in detail. To do this, scientists use what they call active-source seismics — creating booms, then tracing vibrations through the earth with instruments to detect where subterranean breaks interrupt rock layers. Friendly Seattleites almost always let the scientists dig up their lawn to bury a seismometer and let them hook it to their electricity. Some neighbors even compete to land one of the instruments, out of what USGS geophysicist Tom Pratt calls "seismometer envy."

To create the vibrations, scientists have used air guns, shotguns, sledgehammers, explosives, and "thumpers" — pile driver–type trucks that pound the ground with enough force to rattle dishes. (A few years ago scientists had to apologize in the morning paper after one nighttime blast alarmed residents, who thought it was an earthquake.) The USGS also made the most of the city's demolition of its aging Kingdome stadium with explosives in 2000. "We said to ourselves: 'Hey, that's gonna make a big boom!'" says Pratt, who helped plant two hundred seismometers to monitor the event.

One day Atwater and USGS geologist Ray Wells took a ferry to Restoration Point. The flat lower terrace is now a golf course, and people have built expensive homes on the cliff above. From here the scientists pointed out the invisible path of the fault under Puget Sound toward Seattle, past a ten-mile strip of shipping-container piers, petroleum tank farms, and industrial plants, to the city's passenger ferry docks — the country's busiest. As the fault reaches land, it crosses under the waterfront Alaskan Way Viaduct, a 1950s-vintage raised double-decked highway that almost collapsed in the 2001 Nisqually quake and is guaranteed to pancake with anything bigger. (Many geologists avoid driving on it.) Next the fault passes crowds of skyscrapers up to seventy-six stories high and under the two new stadiums housing the Seattle Seahawks football team and the Mariners baseball team. It cuts beneath I-5, proceeds under a steep knoll topped by the headquarters of Amazon.com, forms the southern shoulder of I-90, and heads out to the rapidly growing suburbs around Lake Sammamish.

That is just the Seattle fault; the others zigging across the region could well be connected to it. Many scientists say it is even possible that the faults' activities are connected by some grand mechanism to the great subduction-zone quakes out at sea, for many of the inland quakes seem to have occurred around the same times as those on the sea floor. But the inland mechanics are complicated. According to one currently popular theory, Washington is being pushed by Oregon northward, up against Canada. But Canada is not getting out of the way, so Washington is folding like an accordion, and sometimes those folds — the east-west faults — break violently. "Most people don't want to come right out and say it, but it is all probably linked together in some way we don't understand," says the USGS's Art Frankel.

Geophysicists recently created a stir when they discovered that the deeper part of the ocean slab, subducting from the west under southern British Columbia and northern Washington, slips with uncanny regularity — about every fourteen months — without making conventional seismic waves. No one knows if this "silent" slip relieves tension in the offshore subduction zone or increases it — or if it could somehow help trigger inland quakes. This spring, geophysicists funded by the National Science Foundation will drop instruments into eight deep holes bored into the Olympic Peninsula, west of Seattle, in hopes of monitoring these subtle rumblings. In addition, 150 satellite-controlled global-positioning instruments will be set out across the Northwest to measure minute movements in the crust.

In any event, Seattle is one of the world's worst places for an earthquake. A scenario released last month by a joint private-government group estimates the damage from a 6.7 magnitude shallow crustal quake at $33 billion, with 39,000 buildings largely or totally destroyed, 130 fires burning simultaneously, and 7,700 people dead or badly hurt. Part of the city sits on a soft basin of poorly consolidated sedimentary rocks, and like a bowl of gelatin this unstable base can jiggle if shocked, amplifying seismic waves up to sixteen times. The harbor sits on watery former tidal mud flats, which can liquefy when shaken. One computer model shows a ten-foot tsunami roaring from Puget Sound over the Seattle waterfront to mow down cargo and passenger docks and advancing toward the U.S. Navy shipyards in Bremerton. Even one major bridge collapse would paralyze the city, and engineers predict dozens. Seattle has a lot of high ground — some hillsides are so precipitous that driving up city streets can make one's ears pop — so landslides, already common in heavy rains, are predicted by the thousands.

The city is getting ready, says Ines Pearce, a Seattle emergency manager. A stricter building code was adopted last year. Raised-highway supports are being retrofitted to keep them from crumbling. Firehouse door frames are being reinforced to keep trucks from being trapped inside. Some ten thousand residents have been organized into local disaster-response teams. Schools have removed overhead flush tanks and other hazards, and students duck under their desks in monthly "drop, cover, and hold" earthquake exercises reminiscent of 1950s atomic bomb drills. But the preparations may not be enough. Tom Heaton, a California Institute of Technology geophysicist who first theorized the subduction threat to the Pacific Northwest and is now analyzing Seattle's infrastructure, says that even resistant structures may not survive a major crustal quake or one from the subduction zone. "Earthquake engineers base their designs on past mistakes. No one's ever seen ground shaking like what would occur in a giant earthquake," he says.

Down in the basement of his home, on a leafy Seattle street, Brian Atwater pointed out where he spent $2,000 in the 1990s to reinforce his wooden house frame and bolt it to the concrete foundation to better secure it. During the Nisqually quake, cracks broke out all over his plaster walls, and his chimney got twisted and had to be replaced. But the house didn't go anywhere. If something worse comes along, he hopes the reinforcing will allow his family to escape alive and salvage their possessions.

But there are some risks Atwater *is* willing to abide. On the way back from fieldwork one night recently, he was driving toward his house when he swung his pickup truck away from I-5 — the obvious route — onto the dreaded Alaskan Way Viaduct. Wasn't he nervous? "I'd rather take my chances here," said Atwater, bumping along high over the lights of docks and ships in the harbor. "People over on I-5, they drive too crazy."

KEVIN KRAJICK

The Mummy Doctor

FROM *The New Yorker*

AT THE END OF A STEEP, potholed lane in Duluth, Minnesota, is a three-story brick former elementary school that sits atop a knoll, nearly hidden by a grove of apple and chokecherry trees. A creek flows beneath the building's foundation, through a stone archway. Raspberries and goldenrod edge the windows of the old kindergarten classroom on the ground floor. These days the building houses an assortment of University of Minnesota science labs. The kindergarten room is the headquarters of Dr. Arthur Aufderheide, one of the world's leading experts on the dissection of mummies and a founder of modern paleopathology — the study of ancient diseases.

Aufderheide is a tall, straight-backed man of eighty-two who habitually wears a bolo tie and jumps easily from boulder to boulder on a hike. He has the kindness and good humor of a Midwestern country doctor. The day I met Aufderheide in Duluth, he was wearing a brown tweed jacket and a wash-and-wear pullover shirt whose breast pocket bulged with a black leather notebook — his daily uniform. The shirts, he explained, are "designed to be worn without a T-shirt, so you spend less time getting dressed." He led me past a worn marble staircase into his lab, which houses the International Mummy Registry: six thousand withered chunks of liver, lung, brain, and other tissues that he has taken from some six hundred mummies, from the Arctic to the Sahara.

Upon taking over the space, he had blacked out the wall of windows to keep out the summer sun; the room was now illuminated with cool fluorescent light. In one corner, where an old toy chest used to be, sat a rusting, unplugged 1950s Frigidaire — one of several deceased appliances that provided airtight storage for mummy parts. "I have never cared to possess an entire mummy," Aufderheide said. "It's a formidable challenge to preserve. Even in museums, they often end up getting eaten by insects and mold."

A motley collection of shelves and cabinets contained boxes with labels such as CEMETERY SOIL SAMPLES and MARC KELLEY'S RIBS. There were countless books on medicine and on the ancient world, including the works of Diodorus Siculus and Herodotus. "All knowledge is connected to all other knowledge," Aufderheide said. "The fun is in making the connections."

Paleopathologists value the preserved tissues of mummies because they may harbor signs of ailments or even traces of ancient pathogens. This type of evidence allows scientists to track the historical arc of diseases such as influenza or tuberculosis and determine the cause of medieval plagues like the Black Death. Aufderheide wonders, for example, why cancer seems rare in mummies, no matter where they come from. Is it because these individuals weren't exposed to industrial chemicals, or is there some other factor at work? "Since this is such a new area, almost anything I do is useful," he said.

Aufderheide has written what is considered the authoritative guide to the field, *The Scientific Study of Mummies,* published in 2003 by Cambridge University Press. It includes details such as the nineteen-step recipe used by the Jivaro tribe of Ecuador to shrink the head of a rival killed in war (Step 2: "Transect the neck at the level of the clavicle, pass a band through the mouth and out of the neck, attach a cord, and flee the raided village") and illustrations showing how to explore the innards of someone who has been dead for several thousand years.

"Compared to modern bodies, dissecting mummies is salvage pathology," Aufderheide said. "You lose a great deal. DNA falls apart. Often you're working with an alphabet soup of broken-down proteins where there used to be organs. But we're slowly adapting. The little bit we've done so far shows that we are capable of generating an enormous range of information."

Aufderheide opened the Frigidaire and pulled out a pile of cardboard dog-food boxes, which he had reinforced with duct tape.

Neatly lined up inside them were rows of Whirl-Paks, six-ounce sterile plastic bags designed for taking milk samples at dairies, each containing mummy parts. Some observers say that mummies smell like old books; others say dried leather. But to me they smelled like some dried edible giant ants I had once bought in South America and retained for some years past their prime. It was a faintly acrid, dried-cheese aura — the distilled essence of testy old proteins that had gotten this far and developed a few infirmities but were too mean to give up just yet.

When the human body expires, it usually disappears on a quick, predictable schedule. Within minutes, cell organelles rupture, releasing enzymes that eat the surrounding flesh. Bacteria that inhabit the gut proliferate, race through the visceral veins to the lungs and the heart, then spread to other organs through the arteries. The corpse begins rotting, a process that typically ends, Aufderheide explained, with the "dissolution of skeletal tissue by interaction of bone mineral with ions in the groundwater."

Occasionally, man or nature produces the happy exception that is a mummy. The chief agent here is desiccation; flesh-eating enzymes need water in order to work — "a simple truth exploited by those who bring us beef jerky and dried fruit," Aufderheide said. Thus there are countless natural mummies in places like the Atacama Desert of South America. There, at least seven thousand years ago, the Chinchorro people learned to accelerate the mummification process by placing smoking coals inside their relatives' body cavities after disemboweling them. But the Egyptians, who began making mummies three thousand years later, perfected the art. They systematically removed internal organs (a procedure that eliminates both water and decay-causing microbes), then covered the remains with salts to leech out any remaining moisture, and smeared the corpse with myrrh, pine pitch, and other resinous sealants.

Wrapping the body in linen, or perhaps even just leaving it clothed, seems to contribute to preservation in dry climates, because the material draws moisture away from the body, but it is not necessary. Prehistoric Aleutian Islanders used caves heated by volcanic vents to dry the bodies of their dead, then wrapped them in bird skins and furs; catacombs beneath the sixteenth-century Capuchin monastery in Palermo, Sicily, today hold two thousand unwrapped (though in many cases clothed) mummies, preserved in part by dry limestone walls and excellent ventilation. Frozen mummies are the rarest kind, but occasionally specimens like Ötzi, the so-called Iceman of the Alps, are found embedded in glaciers, mountaintops, or permafrost. Scientists speculate that the peat in European bogs makes mummies by removing metal ions that facilitate decay.

As for more modern methods of mummification, Aufderheide told me that the effect of today's formaldehyde solutions doesn't last that long beyond the funeral. "Formalin will eventually evaporate, and then we're back to the decay process," he said. "I give today's bodies five to thirty years." Aufderheide plans to be cremated when he dies.

Aufderheide slid open a metal drawer. Inside were a few standard scalpels, which he rarely uses, because the flimsy blades often break on time-hardened mummy flesh. Up on a shelf was his Martin autopsy saw, an electric tool with a crescent-shaped, nickel-plated steel blade that efficiently penetrates mummies. Many Third World field sites lack electricity, however, requiring a more primitive approach. From a pants pocket Aufderheide produced his basic road tool, a Swiss Army knife, whose blade he hones razor-sharp with a whetstone that he always carries. Inside the drawer were some curved scissors, good for plucking out eyes. A hacksaw and a chisel, he explained, handle the toughest mummified materials, such as bone.

A mummy's skin, ribs, and chest membranes are generally stuck together. A U-shaped incision through the rib cage, then a line connecting the top of the U below the collarbone, forms a plate that can be lifted like the lid of a hatbox, Aufderheide said. If the mummy is curled up, which is often the case, you can try prying the thighs from the chest with chiropractic firmness. If the mummy resists straightening, you can remove the legs by severing the hip ligaments and twisting counterclockwise; they will have to come off anyway, if this is to be a full dissection. Before lifting the chest plate, you must carefully slide a thin, long-bladed kitchen knife under the edges, to loosen any underlying adhering tissue — otherwise, you might tear out the lungs or heart by accident.

A mummy's interior presents special challenges. The body cavities of the recently deceased are packed with organs, of varying color and consistency. "The liver is usually obvious," Aufderheide said. "It's dark brown and big, and you can follow the hitch down there to the kidney." The lungs are still puffy and pinkish; other organs run to various shades of gray. With a mummy, nearly everything has turned brown and shrunk considerably. Organs lose shape, turn powdery, or go missing. Lungs collapse like old accordions onto the back of the rib cage. Through the centuries, someone's gall bladder could wind up in his pelvis, or his liver in his chest — a phenomenon for which Aufderheide has coined the term "organ migration."

Well-preserved organs may manifest symptoms of disease, but this is rare. Aufderheide normally takes a sample of every tissue present inside a mummy, whatever its apparent state. A thumb-size chunk is enough for lab tests that could reveal hidden germs. Aufderheide estimates that only 10 to 15 percent of mummies show a cause of death; fortunately, chronic ailments reveal far more about people. Parasitic worms of many species are often spotted in coprolites — ancient excrement — in the lower digestive tract. Intestines also host intact bits of ingested vegetables, insect parts, and pollen, which shed light on ancient diets, pests, and the season of the year in which a person died.

As Aufderheide rummaged through his collection, we occasionally encountered objects that were eerily recognizable: a shrunken tongue, speckled with taste buds; a hairy, half-dollar-size ear; ten emaciated white fingers, curled like crab claws, taken from ten Chilean mummies. ("I was working with a rheumatologist at the time," Aufderheide said of the fingers.) He has analyzed only a small percentage of them; as a pioneer in the field of paleopathology, he told me, one of his most important tasks is to carefully extract, catalogue, and organize samples that can one day be studied by others, using techniques that are now still in their infancy.

One box contained a pair of prehistoric Chilean eyeballs. Aufderheide has developed a system of "organ scores" — an index of how often tissues survive — and eyeballs are No. 1. They exist in 93 percent of mummies that have heads. The water inside the eyes drains fast, leaving the durable protein casing and other parts quite intact. The Chilean eyeballs were yellowish-brown, and, with the optic nerves trailing behind, the shriveled orbs resembled wild chanterelles, though others in the collection had collapsed inward so neatly that they looked more like golf tees. "I have scores of eyes, if somebody wants to study them," Aufderheide told me. They could be examined, he said, for signs of eye disease and other ailments, such as hypertension and diabetes, that can be signaled by changes in the eye. "I'm waiting to find someone who's really committed to eyes before I give these away," he said.

Human hair, which is made of tough, insoluble keratin, is just as durable as the eyes. Aufderheide has seen plenty of skeletons that still had lush tresses on their skulls. He pulled out a packet of black hair and a photograph of the ancient Chilean woman to whom it belonged. Her hair was combed and braided, just as loving relatives must have left it. "I often think about what these people were like," he said. "I would like to know what she did every day. How she took care of her kids."

At the bottom of one shelf, in a box that looked as if it were meant to hold greeting cards, were nineteen penises. (Penises are much hardier than, say, breasts, which tend to flatten and disappear.) Many of the penises resembled vacated butterfly cocoons. "I would not know what most of these things were unless I knew where they came from," Aufderheide confessed. He has saved the penises because they "are sponges for blood"; tests on a mummy's preserved blood could, potentially, identify antigens or antibodies linked to specific diseases.

The Whirl-Paks contained many samples of brain matter — pinches of reddish brown granules typical of the brain, which liquefies within days, leaving behind this precipitate when the skull dries out. "If you wanted to study Alzheimer's, you'd be out of luck — you'd need a whole brain," Aufderheide said. But the residue could come in handy someday. "There are some diseases that deposit minerals in the brain," he said, citing Wilson's disease, a rare genetic defect that causes a buildup of copper in the brain. Another set of bags contained mummified bowels, which looked like sausage casings. One sample, which was still flexible, looked grotesquely overstuffed with something — possibly coprolites. Aufderheide palpated the specimen rapidly through the plastic with both hands. Its distended size, he said, could signal Chagas' disease, which is caused by a deadly South American parasite; the disease can kill you by making it impossible for you to defecate.

Aufderheide then extracted a special prize from the refrigerator: a dozen bags containing parts of Acha Man, perhaps the oldest mummy in the world. After Chilean archeologists found Acha Man under an ancient house in the Atacama in 1992, Aufderheide took samples and had him radiocarbon-dated. The results showed that he was about nine thousand years old. "I avoid saying he's *the* oldest mummy," he said. "The moment you say 'oldest,' people will come out of the woodwork and say theirs is a few years older." (Radiocarbon dating has a margin of error running from decades to centuries.) The mummy was missing much of the lower third of his body and most of his organs, but parts of his muscle, skin, and brain remained. Acha Man died young, of unknown causes, though scientists could tell from the thickness of his inner skull that he had probably suffered from chronic iron deficiency. Aufderheide handed me the bag with Acha Man's eyes. They were now just a small, crumbling handful of red and brown debris.

I asked Aufderheide if he wasn't afraid of exhuming a lethal ancient germ. He said no. Bits and pieces of genetic material may survive, but a live ancient pathogen has never been found in a mummy. "They just don't last," he said. Even researchers looking for samples of the 1918 flu strain in Arctic graves several years ago came up with only fragments of the virus. Sometimes Aufderheide wears rubber gloves during dissections, but this is only to avoid getting cut on sharp mummy parts. He said, "I wash my hands after touching a mummy, but really, I'm just going through the motions."

Aufderheide spent decades as a hospital pathologist and part-time medical examiner performing autopsies on what he calls "live people" — the dead victims of Duluth car accidents, strokes, and cancers — before fulfilling a late-in-life desire to study people who lived long ago and far away. He was born in 1922, in New Ulm, Minnesota, a prairie farm town settled by immigrants. His parents, who spoke German at home, ran a brickyard and belonged to a strict Lutheran sect. Aufderheide spent a lot of time in the woods trapping muskrats and dissecting frogs, and he dreamed of being an Arctic explorer. When he expressed an interest in medicine, a teacher told him that if he went to college he hoped he would not be taken in by that nonsense about evolution.

Aufderheide attended medical school during World War II, was inducted into the army, and married Mary Buryk, a nurse from a Ukrainian immigrant family. He served in occupied Europe, and while he was in Vienna he gravitated toward pathology at the Allgemeine Krankenhaus, the sprawling hospital where nineteenth-century doctors had developed the art of the modern autopsy. "With a living person, you treat them, but they go on to die of something else," he said. "You never know how the story turns out. If the person is dead, you know — all the information is in front of you." Cutting into the dead, even babies, didn't faze him. "Many people identify with bodies, which I understand, but I do not. To me, the body is a vehicle that is obviously important in life, but the essence of a person leaves when you die." He has the added advantage of a congenitally deficient sense of smell. "Some people say I'm designed for dissection," he told me, smiling.

In 1953 the Aufderheides moved to Duluth, where Arthur had been offered a post at St. Mary's Hospital. At the time Duluth had no full-time medical examiner, so Aufderheide covered that duty as well. He loved the work, but in his midforties, he told me, "I found myself not as excited to get breakfast down so I could get to work that morning." He and Mary sold their big house and moved, with their daughter and two sons, to a matchbox ranch — probably the smallest physician-owned house in town — so that they could save money for adventure travel. (One son, Tom, is a doctor who specializes in emergency medicine; the other, Walter, is a high school biology teacher. Their sister, Patricia, is a professor of mass communication at American University, in Washington, D.C. She told me that her father used to take her and her brothers to the morgue with him when she was a child but that "it wasn't creepy" when he told her, for example, why a dead old woman's toes had turned blue — it was just another interesting fact about the world.)

In 1964 Aufderheide and three friends floated eleven hundred miles down the Mackenzie River, in Canada, to the Arctic Ocean. Throughout the sixties and seventies, he traveled with Canadian Inuit hunters by dogsled. One day he had a beer with Ralph Plaisted, a Minnesota insurance man and snowmobile enthusiast. Aufderheide idly suggested snowmobiling to the North Pole, and Plaisted organized a trip with the help of corporate sponsors. Aufderheide went along as an amateur navigator. Blizzards and melting sea ice forced them to turn back, however, and the following year Aufderheide gave his spot in the expedition to a professional navigator. Plaisted reached the pole on April 19, 1968, in what was the first successful overland trip since Peary's, in 1909.

When the University of Minnesota opened a new medical school in Duluth in 1972, Aufderheide was hired to teach pathology, a position that required him to pursue academic research. "I was in my fifties," he recalled. "I couldn't see spending seven or eight years getting a Ph.D. in some basic science. So I decided to go with what I had. I'm a physician. I enjoy travel and unfamiliar cultures. Why not study diseases of ancient people?"

In the nineteenth century, when Egyptomania swept Europe and America for the first time, tomb looting yielded large quantities of mummies. There were so many that some were ground up for paint pigments; others were harvested for patent medicines. (In *Innocents Abroad,* Mark Twain dubiously asserts that mummies fueled Egyptian steam locomotives.) Intact specimens appeared in private collections, circuses, and "unwrappings" — quasi-academic theatrical events in which the dead were stripped of their linens to the accompaniment of a lecture or a piano player. Sometimes at unwrappings, physicians did a bit of dissecting. In 1825 Augustus Bozzi Granville, a British surgeon, spotted in the abdomen of one ancient woman a huge ovarian cyst, which he meticulously drew and documented.

More systematic work on mummies started around 1900, just before the partial flooding of the Nile Valley by the first Aswan Dam, as workers disinterred tens of thousands of ancient Egyptians. By then X-rays had been invented and the field of microbiology had been born. While inspecting the kidneys of two three-thousand-year-old mummies, in 1910, Marc Armand Ruffer, a French microbiologist, found dried eggs of the schistosomiasis worm, a parasite that is still prevalent in Egypt. In other mummies he found gallstones, inflamed intestines, and a spleen that had apparently been enlarged by malaria. The insides of many of the ancients' blood vessels, he discovered with surprise, contained calcified spots — hardening of the arteries — even though the Egyptians ate a nutritious low-fat diet centered on cereals. These observations, which suggested that many diseases associated with modernity have a long history, often raised more questions than they answered. Ruffer invented a solution of salts to rehydrate ancient tissues (some researchers today use fabric softener instead) and popularized the word "paleopathology."

The new field lost momentum in the 1920s when anthropologists demonstrated that much information can be gleaned from bones, which are far more plentiful than mummies. Leprosy, syphilis, and tuberculosis attack the eye sockets, the spine, and other skeletal parts in characteristic ways. With proper training, a scientist can look at ancient bones and determine whether the dead person had arthritis or vitamin deficiencies (and say whether he died from a wound inflicted by an ax or a blunt instrument). Even so, some 80 percent of ailments — plague, aneurysms, measles, and many more — leave no marks on bone.

In the early seventies, researchers began to reexamine the mummies exhumed in the nineteenth century with blood-antigen tests, electron microscopy, and other new techniques. Aufderheide knew none of the scientists who were doing this work, but he was determined to break into their circle. He spent several years writing letters to museums around the country, asking for access to their mummies, but he never got a positive response; they didn't want someone with a scalpel anywhere near their collections. While building up his mummy credentials, he studied bones and read esoteric archaeological texts.

Aufderheide learned that some historians attributed the madness of Caligula, and even the decline of the Roman Empire, to poisoning from lead water pipes and wine additives but that no one had ever proved these theories. Starting in the seventies, Aufderheide began following this trail, enlisting Lorentz Wittmers, a university physiologist, to design a new test that could detect lead in bone and eventually joining George Rapp, an archaeologist at his university, in an effort to obtain bone bits collected from twenty Roman-era sites. This led, in 1992, to an article by Aufderheide and his colleagues in the *International Journal of Anthropology* that demonstrated conclusively that the Romans had ten times more lead in their bones than modern Duluthians did.

As his lead studies got under way, Aufderheide began attending whatever relevant scientific meetings he could find and gradually became a presence in the field. In 1978, when Aufderheide was fifty-six, Michael Zimmerman, a pathologist at the University of

Michigan, invited him to a double mummy dissection at Harvard University's Peabody Museum — a sort of modern unwrapping.

When Aufderheide arrived at Harvard, he saw, lying on two lab tables, the bodies of a young thousand-year-old Peruvian man and an elderly Aleut woman from the eighteenth century dressed in an ankle-length robe of sea-otter pelts. Gathered around them were twenty-five people, including physicians, reporters, photographers, and such scientific luminaries as Richard Evans Schultes, the Harvard ethnobotanist who was famous for collecting, and sampling, medicinal plants and hallucinogens acquired from Amazonian shamans. (Schultes wanted to know what plants the mummies' intestines held.) Aufderheide's role, he recalled, "was to stand by with my little drill, to drill some bones for lead."

The mummies were so dried and insubstantial that a grown person could easily have carried them both, one under each arm. The Peruvian man, who was curled in a fetal position, was "only moderately preserved," Aufderheide said. He watched as Zimmerman pulled a vibrating saw from a briefcase, then began cutting through the parched skin of the woman's chest. This procedure spewed a fine brown dust that settled on Aufderheide's oversized glasses and everything else. The doctors lifted the chest plate off and peered in. Inside was a lot of empty space. Then Aufderheide saw the mummy's heart, kidneys, and other organs, all of which had collapsed to the thickness and consistency of corrugated cardboard. "I don't know if I've ever been that excited by a body," he said. The scientists broke off pieces of the organs and passed them around. Aufderheide got some bone, muscle, heart, and intestine to take home.

The dissection revealed that the young man had survived a head injury: he had been hit with a stone from a slingshot. He also bore scars from a trephination, the cutting of a nickel-size hole in his skull, perhaps, by a medicine man in an effort to cure him of headaches resulting from his injuries. The woman seemed to be fairly healthy, despite her age — possibly the benefit of the Aleut diet, which was rich in seafood, the researchers speculated. "It reaffirmed my conviction that there had to be medical information in those tissues," Aufderheide told me.

Three years later Zimmerman called Aufderheide to help with another autopsy, this time of the frozen bodies of an Inuit family that had just been exhumed in Barrow, Alaska — an exceptional opportunity. In the Arctic, bodies are seldom interred more than a foot or two in the permafrost, and frost heaving usually pushes them to the surface, where they soon disintegrate.

Zimmerman and Aufderheide began by prying apart the bodies of two women, which had frozen together. They found that the mummies were remarkably intact, right down to body fluids — which began dripping as the men thawed sections of the bodies with a hair dryer and began cutting. They found the ribs crushed, the hearts constricted, the chest cavities filled with bloody bubbles. They knew that the corpses had been discovered in a seaside dwelling that appeared to have been smashed; the bodies had been beneath fallen timbers, and one woman had thrown her arm up, as if to ward off a blow. The doctors concluded that the women had been crushed by an *ivu* — a giant chunk of sea ice propelled ashore by a spring storm. Radiocarbon dating indicated that they had lived around the beginning of the sixteenth century. Death had apparently occurred in the early morning: the women's stomachs were empty and their bladders were full. The autopsy also revealed significant information about Inuit health. One woman had scar tissue adhering a lung to her chest wall, which indicated that she had suffered from pneumonia but had recovered; she also had hardened arteries. The other woman's muscles carried trichinosis worms — the result, perhaps, of a raw-meat diet. Dissection of the lungs revealed a heavy film of soot: both women had severe black-lung disease, which, Aufderheide and Zimmerman concluded, must have been caused by living indoors all winter with a smoky seal-oil lamp. It was romantic to suppose that these native people had lived healthful outdoor lives. The mummies proved otherwise.

After a few such lucky finds, Aufderheide knew that he had to change his life. Meaningful studies of ancient epidemiology would require hundreds of samples. "I discovered that the big problem in studying mummies is finding them," he said. "You can't wait for bodies to come to you. You have to go looking for the bodies."

In 1985, having heard of untapped specimens in the Atacama and other regions in South America, he and Mary flew to Bogotá, Colombia. The next day the Aufderheides hired a private guide to help them look for mummies; by evening, they had located sixty — in the museum, in church crypts, at the university. Local anthropologists, unlike their counterparts in North America, were eager to collaborate. When the Aufderheides went shopping for souvenirs at a fancy Bogotá shop, they saw three mannequins draped with replicas of pre-Columbian jewelry; as they stepped closer, they realized that the mannequins were mummies. "I knew we were in the right place," Aufderheide said.

With this trip, Mary became her husband's professional partner. Aufderheide spoke no Spanish; Mary had a talent for languages and had just taken a crash course in Spanish. Stops in Peru, Bolivia, and Argentina yielded more mummies. They had their greatest success around Arica, a small Atacama Desert city in northern Chile, where Marvin Allison, a paleopathologist at Virginia Commonwealth University, had set up his own mummy-dissection operation. Here mummies were akin to rocks in a New England field: renewable nuisances, always coming up underfoot in building foundations, water lines, and roadbeds. A local warehouse contained dozens of mummies — from members of the Chinchorro people, who lived thousands of years ago, to nineteenth-century settlers, who had been naturally preserved in their dry graves. The Chinchorro had painstakingly prepared their dead, including babies, often replacing vulnerable facial flesh with clay, which they painted with eyes and other features. Aufderheide was allowed to autopsy twenty-five mummies and to take home samples.

Mary assisted by scraping muscle and skin off bones, writing up labels, and dropping specimens into Whirl-Paks. "People asked me, 'How can you do that?' but it's just like nursing," she told me. "I was enchanted — every body was different. And it kept us together." The Aufderheides frequently returned to South America, and, as Arthur's expertise and standing grew, he and Mary traveled to Spain and the Canary Islands, Hungary, Turkey, Australia, China, and Russia.

Aufderheide, a natural teacher, enlisted dozens of collaborators and disciples. Conrado Rodríguez-Martin, the director of the Canary Islands' Institute for Bioanthropology, said, "It's not just that Art created a new branch of science. It's his friendship. It's his attitude toward life. To many, he is like a father — to me, for sure." In 1992, the Canary Islands hosted the first World Congress on Mummy Studies. It was largely Aufderheide's idea, a way of bringing together members of a distinctly unorganized field. The congress continues to meet every few years.

Larry Cartmell, a middle-aged pathologist from Ada, Oklahoma, volunteered for a dig at Arica and ended up as one of Aufderheide's many apprentices. Forensic scientists had just developed a test that could pinpoint traces of cocaine in hair, and Aufderheide and Cartmell decided to investigate whether ancient Chileans had taken drugs. Hair tests on dozens of mummies definitively traced the movement of coca-chewing from the Andean highlands to the coasts, where, by the fourth century AD, half the population was chewing the leaves — a contradiction of the earlier historical dogma that only aristocrats had indulged in this habit before the Spanish conquest.

Another follower, Derek Notman, a Minnesota radiologist, thought that he and Aufderheide could learn how to use modern medical imaging on human mummies by mummifying someone themselves. Aufderheide rejected this idea; he knew that the university would object. They settled on a forty-pound dog that was going to be put to sleep anyway. They stuffed the animal with natron, the salt used by the Egyptians; it gave off such a stench that it had to be sealed inside multiple garbage bags. After eight months, they had a pretty good mummy. They were able to use it to confirm that body-scanning machines accurately identified interior features useful for spotting lesions, swellings, and other abnormalities. Many investigators now routinely use computerized tomography (CT) scans to reveal the interior of mummies without dissection.

Such developments in imaging technology have, in recent years, made Aufderheide's work more controversial. Heather Pringle, a journalist who observed Aufderheide in the field, later questioned the practice of shredding bodies that had survived millennia. It "was a kind of sacrilege, like smashing a Ming vase in order to discover exactly how it was made," she wrote in her 2001 book *The Mummy Congress*. Some colleagues feel that CT scans and other modern techniques are ready to superannuate the autopsy. Aufderheide, however, points out that many mummies he has worked on would have been destroyed anyway, either torn up by looters or left to rot in poorly funded museums. And scientifically, he told me, the autopsy remains the "gold standard."

In the late eighties Aufderheide began focusing on the paleopathology of tuberculosis. At the time, most scientists believed that the disease first emerged in Middle Eastern cows and spread to people with the development of agriculture. The orthodox theory was that the disease did not reach the New World until Europeans brought it — an idea that neatly explained why so many indigenous peoples died from TB and other allegedly imported pathogens, such as smallpox and malaria. Yet scientists had spotted bone lesions of the sort caused by TB in pre-Columbian mummies from South America, and sudden waves of deadly TB swept Europe for centuries after the Spanish conquest. This led researchers to wonder whether TB was far more ancient than it was generally believed to be, and whether the New World had sent its own lethal strains to infect the Old, as well as vice versa.

"We still don't understand the way many diseases behave on a grand scale," Aufderheide said. "Why does TB ebb and flow over time? Why do some people die of it and others live? If you know how a disease evolved, you have a much better chance of developing a treatment or vaccine. Being able to test past populations will help us do that." As he points out, we still have no vaccine for TB, and, after declining through most of the twentieth century, the disease is again on the rise. In the nineties, new drug-resistant strains evolved, encouraged by poorly supervised drug therapies and the spread of HIV. Some two million people now die annually of tuberculosis.

A major problem that paleopathologists face is the difficulty of making a definitive diagnosis. Tests for antibodies work only if antibodies are still present in the mummies, but often time has robbed them of these delicate molecules. And a disease can be suggested by visual signs but not necessarily confirmed; tuberculosis, for example, may mark the spine or the lungs, but similar lesions can be caused by other ailments.

In 1985 Svante Pääbo, a Swedish molecular biologist, sequenced fragments of human DNA from an Egyptian mummy. Most researchers saw this as an aid to charting human evolution. Aufderheide saw it as a prelude to charting the DNA of ancient pathogens — which would exponentially advance paleopathologists' ability to confirm diagnoses. This would not be an easy task, however: in addition to being tiny, germs have far fewer genes than humans do.

In 1990, Aufderheide was asked by Jane Buikstra, an archeologist who now teaches at the University of New Mexico, to examine some of the hundreds of bodies being excavated from several ancient cemeteries near Ilo, Peru. With Mary's help he set up shop — a sheet of plywood atop two sawhorses — on the patio of a rural house that had been offered to the researchers by a copper-mining company. He awoke well before dawn every morning to eat a huge breakfast of eggs, bacon, hash browns, and coffee — often his only meal until quitting time — then went to work at the first hint of gray in the sky.

"I set up a sort of Henry Ford–type assembly line," Aufderheide told me. On the left he draped a big blue tarp over some empty boxes, where he posed each mummy for exterior photographs. He told the workers to bring anything that was more than just bones. "About half were largely skeletons with ligaments and hair," he said. "The others had varying degrees of abdominal and thoracic organs." To the right of the blue tarp was the plywood, where Mary waited with the autopsy tools and the printed forms they would use for taking notes and making drawings. A family picture shows her standing over the table, wearing a flowery apron, a baseball cap, and a huge smile. ("We have been so lucky," she has said of their travels.)

Aufderheide started with a thorough visual exam — hairs might harbor ancient lice, scars might suggest old injuries — then began cutting. To speed things along, he lined up a row of small cardboard boxes, into which excised organs were dropped. Each box was designated for a different body part: hair, heart, lungs, liver, spleen, intestines, bladder, brain. Following each dissection, Aufderheide took photographs of the corpse's opened torso, then called for the workers to take the dissected mummy away, so that he could move on to the next one. This went on day after day in the unblinking sun. Aufderheide wore a big floppy hat to protect his fair skin, but the hat got in his way, so he often took it off and got blistered. He stopped as the last rays of sun faded behind the horizon. "I had to," he said. "There was no good electric illumination."

One day Aufderheide removed the chest of a middle-aged woman who had died about a thousand years ago and noticed, clinging to her right lung and the attached lymph nodes, a hollow, calcified nodule about the diameter of a dime. He said to himself, "I bet that's Ghon's complex" — a kind of armor that the body builds around tissue killed by TB bacteria in an effort to keep the germ from spreading. He carefully cut out the nodule.

Aufderheide brought the nodule to Wilmar Salo, a biochemist at the university in Duluth. After several false starts — DNA testing was still in its early days — Salo identified in the nodule ninety-seven base pairs of DNA that are peculiar to the human TB germ. In 1994 they published their findings in the *Proceedings of the National Academy of Sciences,* in an article called "Identification of Mycobacterium Tuberculosis DNA in a Pre-Columbian Peruvian Mummy." It was the first time that anyone had conclusively sequenced the DNA of an ancient pathogen — and the last time that scientists questioned the antiquity of TB.

Subsequently modified hypotheses suggest that TB's ancient reservoirs might include mice, llamas, and turkeys. Researchers working on other tissues from Hungary to Borneo have since found TB going back at least 5,400 years. Aufderheide later supplied the mummy samples that resulted in a Colombian team's 1997 identification of ancient DNA from *Trypanosoma cruzi,* the parasite that causes Chagas' disease — the second pathogen so spotted. Other researchers have found fragments of malaria, leprosy, and bubonic plague in ancient Africa and Europe.

For most of his career Aufderheide avoided working on Egyptian mummies because of the Egyptians' habit of removing everything except the heart, which was thought to be the seat of the soul. But in 1993 he took part in a dig in the Dakhla Oasis in the remote Sahara, after seemingly intact mummies were discovered there, and five years later he returned to investigate further. He was now seventy-six years old, three decades older than most of the other scientists there. Dakhla was situated on a forlorn plain of ruined cities and cemeteries, under a great sweep of red sand. Each morning Aufderheide was dropped off at Ismant el-Kharab, known in ancient times as Kellis, which was occupied until the beginning of the fifth century AD. Many small tombs had been dug into a sandstone rock face. He crawled into the tombs and carried out forty-pound mummies, placing them on a couple of planks by a ruined mudbrick wall. The tombs were hiding places for cobras, and blackflies tormented him, but Dakhla had an eerie beauty that moved him.

Aufderheide and his fellow archeologists discovered that some mummies were actually composites of three or four people, which had apparently been reassembled from spare parts after ancient robbers raided the tombs for jewelry. The real treasure for Aufderheide was still there: livers, intestines, kidneys. He forwarded samples of these organs to the newly founded International Ancient Egyptian Mummy Tissue Bank, at the University of Manchester, England, which until then had practically no organ samples.

Toward the close of the Dakhla dig, Aufderheide announced that this would be his last field season. Mary was tired of extreme travel and had declined to accompany him. One afternoon he performed his final dissection. "It was an emotional time, that last mummy," he recalled. "I'd started something from scratch and watched it evolve into something useful. They asked if I wanted someone to come out and assist me. I said, 'No, I'd rather be out there alone.'"

The next morning, stratus clouds spanned the horizon as he watched the sun rise over the mausoleums. He thought about the Egyptian goddess Nut, the mother of the sun, and he thought about death. "I believe in the essence of a person while alive," he said. "But whether it survives I have grave reservations." Before he left, the field crew presented him with a yellow going-away cake in the shape of a mummy.

Today, Aufderheide focuses on his Duluth collection. He and his colleagues are searching for unique proteins secreted by infectious agents like the malaria parasite — molecules that may survive the ravages of time better than DNA, he believes, and that may open new avenues of scientific investigation. He is considering a study of ancient leishmaniasis, a parasitic protozoal disease that eats its victims' skin and mucosal tissues. And he is writing an anthropological book about overmodeled skulls, similar to those of the Chinchorro, that have ritually been supplemented with clay, wood, or other materials; they turn up in diverse cultures around the world. Aufderheide skied cross-country and canoed until two years ago, but he says that he now has little time for recreation. "I have no ambitions beyond what I'm doing," he told me.

Last summer Aufderheide received a call from his friend Gino Fornaciari, a paleopathologist at the University of Pisa. Fornaciari has helped inspire a European mania for exhuming the remains of antique celebrities, including Petrarch and Giotto. (Such work has, for example, identified a cancer-causing gene mutation in a colon tumor that killed Ferdinand of Aragon, the king of Naples, in

1494.) Fornaciari asked Aufderheide to participate in the exhumation of forty-nine members of the Medici family from the family chapel in Florence. The team hopes to identify the diseases that afflicted these Renaissance rulers. It also seeks to confirm or disprove whether certain members of the notoriously violent clan were stabbed, poisoned, or bludgeoned to death, as history has alleged. (Grand Duke Cosimo I had one of many would-be assassins lacerated with red-hot pincers, dragged through Florence's streets by his ankles, and disemboweled.)

This past summer, after some squabbling among modern Medici descendants, Fornaciari and his colleagues pulled up desk-size marble slabs from the chapel floor and removed the first seven bodies. Soon afterward Fornaciari sent Aufderheide a package containing three vertically sawed quarter-sections of vertebrae: one each from Eleonora de Toledo, Giovanni de Medici, and Garzia de Medici, the wife and two sons of Grand Duke Cosimo I. According to some accounts, they died when Garzia stabbed Giovanni, and Cosimo then stabbed Garzia; Eleonora was said to have died of grief. However, the bodies — reduced to bones — showed no signs of violence. Aufderheide and Wilmar Salo, the Duluth biochemist, were examining the vertebrae for malaria germs, which may have been common in Renaissance Italy, and, according to contemporary documents, might actually have killed all three. Even though these were not mummy parts, Aufderheide was excited by them; marrow is rich in blood, and blood carries malaria, which might give him a chance to test his new procedure for finding the proteins the parasite secretes.

During my visit to his lab, Aufderheide retrieved the Medici vertebrae, which he kept inside an old one-gallon pickle jar, along with twelve Whirl-Paks of four-thousand-year-old coprolites from Chile. The fragments of Giovanni, Garzia, and Eleonora were in plastic test tubes: they were mellow to dark brown, fingertip-size chunks, neatly sawed to expose the spongy marrow inside.

Aufderheide held up Giovanni's bone and rattled it around in its tube. I asked him if he got any special feeling from holding part of a historic personage. He winced slightly. "I know I'm supposed to say yes, but this is not the thing that excites me," he said. "I look at this, and I think about the problems of analysis." As for their lives and deaths, "I'm just getting acquainted with this family," Aufderheide said. "Maybe they weren't as bad as people say."

ROBERT KUNZIG

X-Ray Vision

FROM *Discover*

THERE IS A LOT TO SEE IN THE NIGHT SKY, but there is even more not seen. Our eyes have evolved to detect radiation within a narrow range of wavelengths that we call visible light. In essence, we peer out through a slit — and our view is what we used to think was the universe. Now we know better. The real drama happens in places where matter reaches temperatures of millions of degrees and shines mostly in X-rays that our eyes cannot detect. The X-ray sky crackles with previously unimagined action: exploding stars, gas swirling into monster black holes, and pile-driver smashups of whole clusters of galaxies. All of this commotion is finally snapping into focus because of an extraordinary satellite, the Chandra X-ray Observatory, launched by NASA in 1999.

This perspective wouldn't exist if not for a gung ho young physicist named Riccardo Giacconi, who in 1962 talked the air force into letting him launch a Geiger counter into space. NASA, then a brash young agency, had refused to do so. But Giacconi and his team at American Science and Engineering in Cambridge, Massachusetts, already had a contract with the air force to monitor atmospheric nuclear tests, and he knew the air force was hoping to get in on President Kennedy's lunar program. He argued that his Geiger counter might detect X-rays from the moon and thus help determine its composition. "It was a good excuse," he says now.

On June 18, 1962, Giacconi's Geiger counter lifted off on an Aerobee rocket from the White Sands testing range in New Mexico. During the 350 seconds it spent above Earth's X-ray-blocking atmosphere, it registered no emission from the moon but picked up an intense unknown source in the constellation Scorpius. This was the first such source discovered and hence named Scorpius X-1. "Sco X-1 was a boomer," Giacconi recalls. "We had no trouble detecting it."

That result was a happy surprise. Researchers at the Naval Research Laboratory had previously detected X-rays from the sun's hot outer atmosphere, but those rays were only one-millionth as intense as the sun's light. Detecting X-rays from another star light-years away seemed like a long shot. It turned out that Sco X-1 was no ordinary star. It was thousands of times as luminous as the sun, and almost all that radiation was X-rays. "The great thing nature did for us is it invented a brand-new class of stars that nobody expected," Giacconi says. When the air force realized he was looking at distant stars rather than at the moon, they ended his program. By then, however, X-ray astronomy had caught on.

The first scans of the X-ray sky were very coarse. Giacconi's Geiger counter — essentially a box of electrified gas — was fine for recording the passage of X-rays, but it could not create a picture of the source. Even before Sco X-1, however, he had sketched out a design for a true X-ray-imaging telescope. In 1963 he and his colleague Herbert Gursky proposed to NASA a five-year plan that would culminate with a large, orbiting X-ray telescope. Thirty-six years and innumerable bureaucratic snafus later, NASA launched Chandra.

In that time numerous simpler X-ray detectors went up. Giacconi, having worked on one of these experiments, dropped out of the project in 1981, before it had really begun. He passed the leadership of his team at the Harvard-Smithsonian Center for Astrophysics to Harvey Tananbaum, who kept the battles going nearly another two decades. The silver lining is that Chandra is a much better telescope than it would have been if it had been launched with 1960s technology.

Chandra's fundamental design still follows the principles Giacconi sketched in the 1960s. He recognized that you cannot focus X-rays with a lens or reflect them straight off a mirror, because they are so energetic they will burrow right in; that's why X-rays are so good for illuminating the insides of the human body. Giacconi's solution was to direct the X-rays along the sides of a conical mirror, which would cause them to skip along like pebbles off the surface of a pond. To help pull in faint objects, Chandra contains a series of nested mirrors, each of which funnels X-rays to a sharp focus at its narrow end, where a camera sits. And to achieve the desired image clarity — equivalent to reading a stop sign twelve miles away — the mirrors are polished to near perfection, with no bumps larger than six atoms high. In a nice twist of redemption, the job was done successfully by Hughes Danbury Optical Systems, the company that had previously botched the mirror for the Hubble Space Telescope.

What Giacconi could not have foreseen in his original plan, drawn up before the digital revolution, was Chandra's workhorse camera, the ACIS. It detects X-rays using the same kind of silicon chip, called a charge-coupled device, that is in every digital camera. An array of ten chips in ACIS gives it the unique ability to measure both the position and the energy of the incoming rays with high accuracy. Since each element radiates X-rays at a characteristic set of energies, or wavelengths, ACIS can reveal the composition as well as the appearance of objects. "You can make a picture just of silicon. That's the power of Chandra — not just that the mirrors are so good," says astrophysicist Una Hwang of NASA's Goddard Space Flight Center.

So what do these wonderful pictures show? Above all, they expose a universe dominated not by starlight and the fusion energy that creates it but by an energy source no one expected before the discovery of Sco X-1: gravity in its rawest and most extreme forms.

When a massive star, much larger than the sun, exhausts its nuclear fuel, it collapses under its own gravity and then rebounds explosively: that's a supernova. For thousands of years the expanding gases are so hot they emit X-rays. Young supernova remnants, such as Cassiopeia A, are among the most beautiful objects in the X-ray sky. But at the heart of the fireworks lurks an unbeautiful monster, the star's collapsed core.

If the original star was more than ten times as massive as the sun, what's left is an incredibly dense neutron star. If it was more than twenty-five times as massive, the remnant is an even smaller and more bizarre black hole. In the 1960s and 1970s Giacconi's team provided some of the first strong evidence that black holes were not just the fever dreams of theorists. The objects give themselves away by what they do to their neighbors. Thanks to its intense gravity, a neutron star or a black hole in orbit with an ordinary star snatches gas from its companion. That gas, swirling and plunging into the gravitational pit, becomes so hot it emits X-rays — a lot of X-rays. A proton falling into a neutron star releases fifty times as much energy as a proton fusing with another inside the sun. That's why Giacconi's detectors could detect Sco X-1 (now recognized as a neutron-star binary) even though it lies some 1,000 light-years from Earth.

On the grand scale, even Sco X-1 is a pip-squeak. Astronomers now think that the center of our Milky Way is home to a black hole nearly three million times as massive as the sun. With Chandra, they are watching its sputterings in unprecedented detail. Other galaxies harbor even more powerful black holes, billions of times as heavy as the sun. When stars or vast clouds of gas fall into such massive objects, the resulting X-ray blaze can be seen across the universe.

Chandra also observes many other ways that gravity sculpts the cosmos. The potent mutual attraction that holds together huge clusters of galaxies traps hot gas that feverishly emits X-rays. The contours of these emissions trace the collisions and mergers that created these clusters and that continue to tear apart and reshape the individual galaxies within. Galaxies can run headlong into each other, sparking the furious birth of hot, short-lived stars; some of these go on to create the next generation of neutron stars and black holes. Chandra has watched groups of new stars pulling themselves together and has monitored one hugely unstable star, Eta Carinae, which seems on the verge of pulling itself apart.

What Chandra sees, near and far, is nature at its most energetic. "Gone is the classical conception of the universe as a serene and majestic ensemble," Giacconi told the audience in Stockholm in December 2002, when he received a Nobel Prize for his work. "The universe we know today is pervaded by the echoes of enormous explosions and rent by abrupt changes." No other X-ray telescope, existing or planned, provides a sharper look at that action than Chandra. "These are the best pictures we're going to see for a long time," Una Hwang says.

JUAN MALDACENA

The Illusion of Gravity

FROM *Scientific American*

THREE SPATIAL DIMENSIONS are visible all around us — up/down, left/right, forward/backward. Add time to the mix, and the result is a four-dimensional blending of space and time known as spacetime. Thus we live in a four-dimensional universe. Or do we?

Amazingly, some new theories of physics predict that one of the three dimensions of space could be a kind of illusion — that in actuality all the particles and fields that make up reality are moving about in a two-dimensional realm like the Flatland of Edwin A. Abbott. Gravity, too, would be part of the illusion: a force that is not present in the two-dimensional world but that materializes along with the emergence of the illusory third dimension.

More precisely, the theories predict that the number of dimensions in reality could be a matter of perspective: physicists could choose to describe reality as obeying one set of laws (including gravity) in three dimensions or, equivalently, as obeying a different set of laws that operates in two dimensions (in the absence of gravity). Despite the radically different descriptions, both theories would describe everything that we see and all the data we could gather about how the universe works. We would have no way to determine which theory was "really" true.

Such a scenario strains the imagination. Yet an analogous phenomenon occurs in everyday life. A hologram is a two-dimensional object, but when viewed under the correct lighting conditions it produces a fully three-dimensional image. All the information describing the three-dimensional image is in essence encoded in the two-dimensional hologram. Similarly, according to the new physics theories, the entire universe could be a kind of a hologram.

The holographic description is more than just an intellectual or philosophical curiosity. A computation that might be very difficult in one realm can turn out to be relatively straightforward in the other, thereby turning some intractable problems of physics into ones that are easily solved. For example, the theory seems useful in analyzing a recent experimental high-energy physics result. Moreover, the holographic theories offer a fresh way to begin constructing a quantum theory of gravity — a theory of gravity that respects the principles of quantum mechanics. A quantum theory of gravity is a key ingredient in any effort to unify all the forces of nature, and it is needed to explain both what goes on in black holes and what happened in the nanoseconds after the big bang. The holographic theories provide potential resolutions of profound mysteries that have dogged attempts to understand how a theory of quantum gravity could work.

A Difficult Marriage

A quantum theory of gravity is a holy grail for a certain breed of physicist because all of physics, except for gravity, is well described by quantum laws. The quantum description of physics represents an entire paradigm for physical theories, and it makes no sense for one theory, gravity, to fail to conform to it. Now about eighty years old, quantum mechanics was first developed to describe the behavior of particles and forces in the atomic and subatomic realms. It is at those size scales that quantum effects become significant. In quantum theories, objects do not have definite positions and velocities but instead are described by probabilities and waves that occupy regions of space. In a quantum world, at the most fundamental level everything is in a state of constant flux, even "empty" space, which is in fact filled with virtual particles that perpetually pop in and out of existence.

In contrast, physicists' best theory of gravity, general relativity, is an inherently classical (that is, nonquantum) theory. Einstein's magnum opus, general relativity explains that concentrations of matter or energy cause spacetime to curve and that this curvature deflects the trajectories of particles, just as should happen for particles in a gravitational field. General relativity is a beautiful theory, and many of its predictions have been tested to great accuracy.

In a classical theory such as general relativity, objects have definite locations and velocities, like the planets orbiting the sun. One can plug those locations and velocities (and the masses of the objects) into the equations of general relativity and deduce the curvature of spacetime and from that deduce the effects of gravity on the objects' trajectories. Furthermore, empty spacetime is perfectly smooth no matter how closely one examines it — a seamless arena in which matter and energy can play out their lives.

The problem in devising a quantum version of general relativity is not just that on the scale of atoms and electrons, particles do not have definite locations and velocities. To make matters worse, at the even tinier scale delineated by the Planck length (10^{-33} centimeter), quantum principles imply that spacetime itself should be a seething foam, similar to the sea of virtual particles that fills empty space. When matter and spacetime are so protean, what do the equations of general relativity predict? The answer is that the equations are no longer adequate. If we assume that matter obeys the laws of quantum mechanics and gravity obeys the laws of general relativity, we end up with mathematical contradictions. A quantum theory of gravity (one that fits within the paradigm of quantum theories) is needed.

In most situations, the contradictory requirements of quantum mechanics and general relativity are not a problem, because either the quantum effects or the gravitational effects are so small that they can be neglected or dealt with by approximations. When the curvature of spacetime is very large, however, the quantum aspects of gravity become significant. It takes a very large mass or a great concentration of mass to produce much spacetime curvature. Even the curvature produced near the sun is exceedingly small compared with the amount needed for quantum gravity effects to become apparent.

Though these effects are completely negligible now, they were very important at the beginning of the big bang, which is why a quantum theory of gravity is needed to describe how the big bang started. Such a theory is also important for understanding what happens at the center of black holes, because matter there is crushed into a region of extremely high curvature. Because gravity involves spacetime curvature, a quantum gravity theory will also be a theory of quantum spacetime; it should clarify what constitutes the "spacetime foam" mentioned earlier, and it will probably provide us with an entirely new perspective on what spacetime is at the deepest level of reality.

A very promising approach to a quantum theory of gravity is string theory, which some theoretical physicists have been exploring since the 1970s. String theory overcomes some of the obstacles to building a logically consistent quantum theory of gravity. String theory, however, is still under construction and is not yet fully understood. That is, we string theorists have some approximate equations for strings, but we do not know the exact equations. We also do not know the guiding underlying principle that explains the form of the equations, and there are innumerable physical quantities that we do not know how to compute from the equations.

In recent years string theorists have obtained many interesting and surprising results, giving novel ways of understanding what a quantum spacetime is like. I will not describe string theory in much detail here but instead will focus on one of the most exciting recent developments emerging from string-theory research, which led to a complete, logically consistent quantum description of gravity in what are called negatively curved spacetimes — the first such description ever developed. For these spacetimes, holographic theories appear to be true.

Negatively Curved Spacetimes

All of us are familiar with Euclidean geometry, in which space is flat (that is, not curved). It is the geometry of figures drawn on flat sheets of paper. To a very good approximation, it is also the geometry of the world around us: parallel lines never meet, and all the rest of Euclid's axioms hold.

We are also familiar with some curved spaces. Curvature comes in two forms, positive and negative. The simplest space with positive curvature is the surface of a sphere, which has constant positive curvature. That is, it has the same degree of curvature at every location (unlike an egg, say, which has more curvature at the pointy end).

The simplest space with negative curvature is called hyperbolic space, which is defined as space with constant negative curvature. This kind of space has long fascinated scientists and artists alike. Indeed, M. C. Escher produced several beautiful pictures of hyperbolic space. One such picture, of fish, is like a flat map of the space; the way that the fish become smaller and smaller is just an artifact of how the curved space is squashed to fit on a flat sheet of paper, similar to the way that countries near the poles get stretched on a map of the globe (a sphere).

By including time in the game, physicists can similarly consider space*times* with positive or negative curvature. The simplest spacetime with positive curvature is called de Sitter space, after Willem de Sitter, the Dutch physicist who introduced it. Many cosmologists believe that the very early universe was close to being a de Sitter space. The far future may also be de Sitter–like because of cosmic acceleration. Conversely, the simplest negatively curved spacetime is called anti–de Sitter space. It is similar to hyperbolic space except that it also contains a time direction. Unlike our universe, which is expanding, anti–de Sitter space is neither expanding nor contracting. It looks the same at all times. Despite that difference, anti–de Sitter space turns out to be quite useful in the quest to form quantum theories of spacetime and gravity.

If we picture hyperbolic space as being a disk like Escher's drawing of fish, then anti–de Sitter space is like a stack of those disks, forming a solid cylinder. Time runs along the cylinder. Hyperbolic space can have more than two spatial dimensions. The anti–de Sitter space most like our spacetime (with three spatial dimensions) would have a three-dimensional "Escher print" as the cross section of its "cylinder."

Physics in anti–de Sitter space has some strange properties. If you were freely floating anywhere in anti–de Sitter space, you would feel as though you were at the bottom of a gravitational well. Any object that you threw out would come back like a boomerang. Surprisingly, the time required for an object to come back would be independent of how hard you threw it. The difference would just be that the harder you threw it, the farther away it would get on its round trip back to you. If you sent a flash of light, which consists of photons moving at the maximum possible speed (the speed of light), it would actually reach infinity and come back to you, all in a finite amount of time. This can happen because an object experiences a kind of time contraction of ever greater magnitude as it gets farther away from you.

The Hologram

Anti–de Sitter space, although it is infinite, has a "boundary," located out at infinity. To draw this boundary, physicists and mathematicians use a distorted length scale similar to Escher's, squeezing an infinite distance into a finite one. This boundary is like the outer circumference of the Escher print or the surface of the solid cylinder I considered earlier. In the cylinder example, the boundary has two dimensions — one is space (looping around the cylinder), and one is time (running along its length). For four-dimensional anti–de Sitter space, the boundary has two space dimensions and one time dimension. Just as the boundary of the Escher print is a circle, the boundary of four-dimensional anti–de Sitter space at any moment in time is a sphere. This boundary is where the hologram of the holographic theory lies.

Stated simply, the idea is as follows: a quantum gravity theory in the interior of an anti–de Sitter spacetime is completely equivalent to an ordinary quantum particle theory living on the boundary. If true, this equivalence means that we can use a quantum particle theory (which is relatively well understood) to define a quantum gravity theory (which is not).

To make an analogy, imagine you have two copies of a movie, one on reels of 70-millimeter film and one on a DVD. The two formats are utterly different, the first a linear ribbon of celluloid with each frame recognizably related to scenes of the movie as we know it, the second a pitted disk that reflects laser light that would form a sequence of 0s and 1s if we could perceive them at all. Yet both "describe" the same movie.

Similarly, the two theories, superficially utterly different in content, describe the same universe. The DVD looks like a metal disk with some glints of rainbowlike patterns. The boundary particle theory "looks like" a theory of particles in the absence of gravity. From the DVD, detailed pictures emerge only when the bits are processed the right way. From the boundary particle theory, quantum gravity and an extra dimension emerge when the equations are analyzed the right way.

What does it really mean for the two theories to be equivalent? First, for every entity in one theory, the other theory has a counterpart. The entities may be very different in how they are described by the theories: one entity in the interior might be a single particle of some type, corresponding on the boundary to a whole collection of particles of another type, considered as one entity. Second, the predictions for corresponding entities must be identical. Thus, if two particles have a 40 percent chance of colliding in the interior, the two corresponding collections of particles on the boundary should also have a 40 percent chance of colliding.

Here is the equivalence in more detail. The particles that live on the boundary interact in a way that is very similar to how quarks and gluons interact in reality (quarks are the constituents of protons and neutrons; gluons generate the strong nuclear force that binds the quarks together). Quarks have a kind of charge that comes in three varieties, called colors, and the interaction is called chromodynamics. The difference between the boundary particles and ordinary quarks and gluons is that the particles have a large number of colors, not just three.

Gerard 't Hooft of Utrecht University in the Netherlands studied such theories as long ago as 1974 and predicted that the gluons would form chains that behave much like the strings of string theory. The precise nature of these strings remained elusive, but in 1981 Alexander M. Polyakov, now at Princeton University, noticed that the strings effectively live in a higher-dimensional space than the gluons do. As we shall see shortly, in our holographic theories that higher-dimensional space is the interior of anti–de Sitter space.

To understand where the extra dimension comes from, start by considering one of the gluon strings on the boundary. This string has a thickness, related to how much its gluons are smeared out in space. When physicists calculate how these strings on the boundary of anti–de Sitter space interact with one another, they get a very odd result: two strings with different thicknesses do not interact very much with each other. It is as though the strings were separated spatially. One can reinterpret the thickness of the string to be a new spatial coordinate that goes away from the boundary.

Thus a thin boundary string is like a string close to the boundary, whereas a thick boundary string is like one far away from the boundary. The extra coordinate is precisely the coordinate needed to describe motion within the four-dimensional anti–de Sitter spacetime! From the perspective of an observer in the spacetime, boundary strings of different thicknesses appear to be strings (all of them thin) at different radial locations. The number of colors on the boundary determines the size of the interior (the radius of the Escher-like sphere). To have a spacetime as large as the visible universe, the theory must have about 10^{60} colors.

It turns out that one type of gluon chain behaves in the four-dimensional spacetime as the graviton, the fundamental quantum particle of gravity. In this description, gravity in four dimensions is an emergent phenomenon arising from particle interactions in a gravityless, three-dimensional world. The presence of gravitons in the theory should come as no surprise — physicists have known since 1974 that string theories always give rise to quantum gravity. The strings formed by gluons are no exception, but the gravity operates in the higher-dimensional space.

Thus the holographic correspondence is not just a wild new possibility for a quantum theory of gravity. Rather, in a fundamental way, it connects string theory, the most studied approach to quantum gravity, with theories of quarks and gluons, which are the cornerstone of particle physics. What is more, the holographic theory seems to provide some insight into the elusive exact equations of string theory. String theory was actually invented in the late 1960s for the purpose of describing strong interactions, but it was later abandoned (for that purpose) when the theory of chromodynamics entered the scene. The correspondence between string theory and chromodynamics implies that these early efforts were not misguided; the two descriptions are different faces of the same coin.

Varying the boundary chromodynamics theory by changing the details of how the boundary particles interact gives rise to an assortment of interior theories. The resulting interior theory can have only gravitational forces, or gravity plus some extra force such as the electromagnetic force, and so on. Unfortunately, we do not yet know of a boundary theory that gives rise to an interior theory that includes exactly the four forces we have in our universe.

I first conjectured that this holographic correspondence might hold for a specific theory (a simplified chromodynamics in a four-dimensional boundary spacetime) in 1997. This immediately excited great interest from the string-theory community. The conjecture was made more precise by Polyakov, Stephen S. Gubser, and Igor R. Klebanov of Princeton University and Edward Witten of the

Institute for Advanced Study in Princeton. Since then, many researchers have contributed to exploring the conjecture and generalizing it to other dimensions and other chromodynamics theories, providing mounting evidence that it is correct. So far, however, no example has been rigorously proved — the mathematics is too difficult.

Mysteries of Black Holes

How does the holographic description of gravity help to explain aspects of black holes? Black holes are predicted to emit Hawking radiation, named after Stephen W. Hawking of the University of Cambridge, who discovered this result. This radiation comes out of the black hole at a specific temperature. For all ordinary physical systems, a theory called statistical mechanics explains temperature in terms of the motion of the microscopic constituents. This theory explains the temperature of a glass of water or the temperature of the sun. What about the temperature of a black hole? To understand it, we would need to know what the microscopic constituents of the black hole are and how they behave. Only a theory of quantum gravity can tell us that.

Some aspects of the thermodynamics of black holes have raised doubts as to whether a quantum-mechanical theory of gravity could be developed at all. It seemed as if quantum mechanics itself might break down in the face of effects taking place in black holes. For a black hole in an anti–de Sitter spacetime, we now know that quantum mechanics remains intact, thanks to the boundary theory. Such a black hole corresponds to a configuration of particles on the boundary. The number of particles is very large, and they are all zipping around, so theorists can apply the usual rules of statistical mechanics to compute the temperature. The result is the same as the temperature that Hawking computed by very different means, indicating that the results can be trusted. Most important, the boundary theory obeys the ordinary rules of quantum mechanics; no inconsistency arises.

Physicists have also used the holographic correspondence in the opposite direction — employing known properties of black holes in the interior spacetime to deduce the behavior of quarks and gluons at very high temperatures on the boundary. Dam Son of the

University of Washington and his collaborators studied a quantity called the shear viscosity, which is small for a fluid that flows very easily and large for a substance more like molasses. They found that black holes have an extremely low shear viscosity — smaller than any known fluid. Because of the holographic equivalence, strongly interacting quarks and gluons at high temperatures should also have very low viscosity.

A test of this prediction comes from the Relativistic Heavy Ion Collider (RHIC) at Brookhaven National Laboratory, which has been colliding gold nuclei at very high energies. A preliminary analysis of these experiments indicates that the collisions are creating a fluid with very low viscosity. Even though Son and his coworkers studied a simplified version of chromodynamics, they seem to have come up with a property that is shared by the real world. Does this mean that RHIC is creating small five-dimensional black holes? It is really too early to tell, both experimentally and theoretically. (Even if so, there is nothing to fear from these tiny black holes — they evaporate almost as fast as they are formed, and they "live" in five dimensions, not in our own four-dimensional world.)

Many questions about the holographic theories remain to be answered. In particular, does anything similar hold for a universe like ours in place of the anti–de Sitter space? A crucial aspect of anti–de Sitter space is that it has a boundary where time is well defined. The boundary has existed and will exist forever. An expanding universe, like ours, that comes from a big bang does not have such a well-behaved boundary. Consequently, it is not clear how to define a holographic theory for our universe; there is no convenient place to put the hologram.

An important lesson that one can draw from the holographic conjecture, however, is that quantum gravity, which has perplexed some of the best minds on the planet for decades, can be very simple when viewed in terms of the right variables. Let's hope we will soon find a simple description for the big bang!

CHARLES C. MANN

The Coming Death Shortage

FROM *The Atlantic Monthly*

ANNA NICOLE SMITH'S role as a harbinger of the future is not widely acknowledged. Born Vickie Lynn Hogan, Smith first came to the attention of the American public in 1993, when she earned the title Playmate of the Year. In 1994 she married J. Howard Marshall, a Houston oil magnate said to be worth more than half a billion dollars. He was eighty-nine and wheelchair-bound; she was twenty-six and quiveringly mobile. Fourteen months later Marshall died. At his funeral the widow appeared in a white dress with a vertical neckline. She also claimed that Marshall had promised half his fortune to her. The inevitable litigation sprawled from Texas to California and occupied batteries of lawyers, consultants, and public relations specialists for more than seven years.

Even before Smith appeared, Marshall had disinherited his older son. And he had infuriated his younger son by lavishing millions on a mistress, an exotic dancer, who then died in a bizarre face-lift accident. To block Marshall senior from squandering on Smith money that Marshall junior regarded as rightfully his, the son seized control of his father's assets by means that the trial judge later said were so "egregious," "malicious," and "fraudulent" that he regretted being unable to fine the younger Marshall more than $44 million in punitive damages.*

* After this article was submitted, a federal appellate court ruled that the Anna Nicole Smith case properly should have been decided not by the federal court in California that awarded her $88 million but by the Texas probate court that had previously ruled wholly against her. Smith appealed to the U.S. Supreme Court, joined by the Bush administration, which wished to preserve federal jurisdiction

In its epic tawdriness the Marshall affair was natural fodder for the tabloid media. Yet one aspect of it may soon seem less a freak show than a cliché. If an increasingly influential group of researchers is correct, the lurid spectacle of intergenerational warfare will become a typical social malady.

The scientists' argument is circuitous but not complex. In the past century U.S. life expectancy has climbed from forty-seven to seventy-seven, increasing by nearly two thirds. Similar rises happened in almost every country. And this process shows no sign of stopping: according to the United Nations, by 2050 global life expectancy will have increased by another ten years. Note, however, that this tremendous increase has been in *average* life expectancy — that is, the number of years that most people live. There has been next to no increase in the *maximum* lifespan, the number of years that one can possibly walk the earth — now thought to be about 120. In the scientists' projections, the ongoing increase in average lifespan is about to be joined by something never before seen in human history: a rise in the maximum possible age at death.

Stem-cell banks, telomerase amplifiers, somatic gene therapy — the list of potential longevity treatments incubating in laboratories is startling. Three years ago a multi-institutional scientific team led by Aubrey de Grey, a theoretical geneticist at Cambridge University, argued in a widely noted paper that the first steps toward "engineered negligible senescence" — a rough-and-ready version of immortality — would have "a good chance of success in mice within ten years." The same techniques, De Grey says, should be ready for human beings a decade or so later. "In ten years we'll have a pill that will give you twenty years," says Leonard Guarente, a professor of biology at Massachusetts Institute of Technology. "And then there'll be another pill after that. The first hundred-and-fifty-year-old may have already been born."

Critics regard such claims as wildly premature. In March ten respected researchers predicted in the *New England Journal of Medicine* that "the steady rise in life expectancy during the past two

---

over state probate proceedings. The court ruled 9–0 for Smith in May 2006, sending the case back to appellate court. No matter who finally wins this *Bleak House* legal battle, though, the case remains emblematic of the social conflicts that will ensue as the interests of increasingly long-lived older generations diverge from those of their heirs.

centuries may soon come to an end," because rising levels of obesity are making people sicker. The research team leader, S. Jay Olshansky, of the University of Illinois School of Public Health, also worries about the "potential impact of infectious disease." Believing that medicine can and will overcome these problems, his "cautious and I think defensibly optimistic estimate" is that the average lifespan will reach eighty-five or ninety — in 2100. Even this relatively slow rate of increase, he says, will radically alter the underpinnings of human existence. "Pushing the outer limits of lifespan" will force the world to confront a situation no society has ever faced before: an acute shortage of dead people.

The twentieth-century jump in life expectancy transformed society. Fifty years ago senior citizens were not a force in electoral politics. Now the AARP is widely said to be the most powerful organization in Washington. Medicare, Social Security, retirement, Alzheimer's, snowbird economies, the population boom, the golfing boom, the cosmetic-surgery boom, the nostalgia boom, the recreational-vehicle boom, Viagra — increasing longevity is entangled in every one. Momentous as these changes have been, though, they will pale before what is coming next.

From religion to real estate, from pensions to parent-child dynamics, almost every aspect of society is based on the orderly succession of generations. Every quarter century or so, children take over from their parents — a transition as fundamental to human existence as the rotation of the planet about its axis. In tomorrow's world, if the optimists are correct, grandparents will have living grandparents; children born decades from now will ignore advice from people who watched the Beatles on *The Ed Sullivan Show.* Intergenerational warfare — the Anna Nicole Smith syndrome — will be but one consequence. Trying to envision such a world, sober social scientists find themselves discussing pregnant seventy-year-olds, offshore organ farms, protracted adolescence, and lifestyles policed by insurance companies. Indeed, if the biologists are right, the coming army of centenarians will be marching into a future so unutterably different that they may well feel nostalgia for the long-ago days of three score and ten.

The oldest in vitro fertilization clinic in China is located on the sixth floor of a no-star hotel in Changsha, a gritty flyover city in the south-central portion of the country. It is here that the clinic's founder and director, Lu Guangxiu, pursues her research into embryonic stem cells.

Most cells *don't* divide, in spite of what elementary school students learn — they just get old and die. The body subcontracts out the job of replacing them to a special class of cells called stem cells. Embryonic stem cells — those in an early-stage embryo — can grow into any kind of cell: spleen, nerve, bone, whatever. Rather than having to wait for a heart transplant, medical researchers believe, a patient could use stem cells to grow a new heart: organ transplant without an organ donor.

The process of extracting stem cells destroys an early-stage embryo, which has led the Bush administration to place so many strictures on stem-cell research that scientists complain it has been effectively banned in this country. A visit to Lu's clinic not long ago suggested that ultimately Bush's rules won't stop anything. Capitalism won't let them.

During a conversation Lu accidentally brushed some papers to the floor. They were faxes from venture capitalists in San Francisco, Hong Kong, and Stuttgart. "I get those all the time," she said. Her operation was short of money — a chronic problem for scientists in poor countries. But it had something of value: thousands of frozen embryos, an inevitable byproduct of in vitro fertilizations. After obtaining permission from patients, Lu uses the embryos in her work. It is possible that she has access to more embryonic stem cells than all U.S. researchers combined.

Sooner or later, in one nation or another, someone like Lu will cut a deal: frozen embryos for financial backing. Few are the stem-cell researchers who believe that their work will not lead to tissue-and-organ farms and that these will not have a dramatic impact on the human lifespan. If Organs 'Я' Us is banned in the United States, Americans will seek out longevity centers elsewhere. As Stephen S. Hall wrote in *Merchants of Immortality,* biotechnology increasingly resembles the software industry. Dependence on venture capital, loathing of regulation, pathological secretiveness, penchant for hype, willingness to work overseas — they're all there. Already the U.S. Patent Office has issued four hundred patents concerning human stem cells.

Longevity treatments will almost certainly drive up medical costs, says Dana Goldman, the director of health economics at the RAND Corporation, and some might drive them up significantly. Implanted defibrillators, for example, could constantly monitor people's hearts for signs of trouble, electrically regulating the organs when they miss a beat. Researchers believe that the devices would reduce heart-disease deaths significantly. At the same time, Goldman says, they would by themselves drive up the nation's health-care costs by "many billions of dollars" (Goldman and his colleagues are working on nailing down how much), and they would be only one of many new medical interventions. In developed nations antiretroviral drugs for AIDS typically cost about $15,000 a year. According to James Lubitz, the acting chief of the aging and chronic-disease statistics branch of the Centers for Disease Control's National Center for Health Statistics, there is no a priori reason to suppose that lifespan extension will be cheaper, that the treatments will have to be administered less frequently, or that their inventors will agree to be less well compensated. To be sure, as Ramez Naam points out in *More Than Human,* which surveys the prospects for "biological enhancement," drugs inevitably fall in price as their patents expire. But the same does not necessarily hold true for medical procedures: heart bypass operations are still costly, decades after their invention. And in any case there will invariably be newer, more effective, and more costly drugs. Simple arithmetic shows that if 80 million U.S. senior citizens were to receive $15,000 worth of treatment every year, the annual cost to the nation would be $1.2 trillion — "the kind of number," Lubitz says, "that gets people's attention."

The potential costs are enormous, but the United States is a rich nation. As a share of gross domestic product, the cost of U.S. health care roughly doubled from 1980 to the present, explains David M. Cutler, a health-care economist at Harvard. Yet unlike many cost increases, this one signifies that people are better off. "Would you rather have a heart attack with 1980 medicine at the 1980 price?" Cutler asks. "We get more and better treatments now, and we pay more for the additional services. I don't look at that and see an obvious disaster."

The critical issue, in Goldman's view, will be not the costs per se but determining who will pay them. "We're going to have a very public debate about whether this will be covered by insurance," he says. "My sense is that it won't. It'll be like cosmetic surgery — you pay out of pocket." Necessarily, a pay-as-you-go policy would limit access to longevity treatments. If high-level antiaging therapy were expensive enough, it could become a perk for movie stars, politicians, and CEOs. One can envision Michael Moore fifty years from now, still denouncing the rich in political tracts delivered through the next generation's version of the Internet — neural implants, perhaps. Donald Trump, a 108-year-old multibillionaire in 2054, will be firing the children of the apprentices he fired in 2004. Meanwhile, the maids, chauffeurs, and gofers of the rich will stare mortality in the face.

Short of overtly confiscating rich people's assets, it would be hard to avoid this divide. Yet as Goldman says, there will be "furious" political pressure to avert the worst inequities. For instance, government might mandate that insurance cover longevity treatments. In fact, it is hard to imagine any democratic government foolhardy enough *not* to guarantee access to those treatments, especially when the old are increasing in number and political clout. But forcing insurers to cover longevity treatments would only change the shape of the social problem. "Most everyone will want to take [the treatment]," Goldman says. "So that jacks up the price of insurance, which leads to more people uninsured. Either way, we may be bifurcating society."

Ultimately, Goldman suggests, the government would probably end up paying outright for longevity treatments — an enormous new entitlement program. How could it be otherwise? Older voters would want it because it is in their interest; younger ones would want it because they, too, will age. "At the same time," he says, "nobody likes paying taxes, so there would be constant pressure to contain costs."

To control spending, the program might give priority to people with healthy habits; no point in retooling the genomes of smokers, risk takers, and addicts of all kinds. A kind of reverse eugenics might occur, in which governments would freely allow the birth of people with "bad" genes but would let nature take its course on them as they aged. Having shed the baggage of depression, addiction, mental retardation, and chemical-sensitivity syndrome, tomorrow's legions of perduring old would be healthier than the young. In this scenario moralists and reformers would have a field day.

Meanwhile, the gerontocratic elite will have a supreme weapon against the young: compound interest. According to a 2004 study by three researchers at the London Business School, historically the average rate of real return on stock markets worldwide has been about 5 percent. Thus a twenty-year-old who puts $10,000 in the market in 2010 should expect by 2030 to have about $27,000 in real terms — a tidy increase. But that happy forty-year-old will be in the same world as septuagenarians and octogenarians who began investing their money during the Carter administration. If someone who turned seventy in 2010 had invested $10,000 when he was twenty, he would have about $115,000. In the same twenty-year period during which the young person's account grew from $10,000 to $27,000, the old person's account would grow from $115,000 to $305,000. Inexorably, the gap between them will widen.

The result would be a tripartite society: the very old and very rich on top, beta-testing each new treatment on themselves; a mass of the ordinary old, forced by insurance into supremely healthy habits, kept alive by medical entitlement; and the diminishingly influential young. In his novel *Holy Fire* (1996) the science fiction writer and futurist Bruce Sterling conjured up a version of this dictatorship-by-actuary: a society in which the cautious, careful centenarian rulers, supremely fit and disproportionately affluent, if a little frail, look down with ennui and mild contempt on their juniors. Marxist class warfare, upgraded to the biotech era!

In the past, twenty- and thirty-year-olds had the chance of sudden windfalls in the form of inheritances. Some economists believe that bequests from previous generations have provided as much as a quarter of the start-up capital for each new one — money for college tuitions, new houses, new businesses. But the image of an ingénue's getting a leg up through a sudden bequest from Aunt Tilly will soon be a relic of late-millennium romances.

Instead of helping their juniors begin careers and families, tomorrow's rich oldsters will be expending their disposable income to enhance their memories, senses, and immune systems. Refashioning their flesh to ever higher levels of performance, they will adjust their metabolisms on computers, install artificial organs that synthesize smart drugs, and swallow genetically tailored bacteria and viruses that clean out arteries, fine-tune neurons, and repair broken genes. Should one be reminded of H. G. Wells's *The Time Machine,* in which humankind is divided into two species, the ethereal Eloi and the brutish, underground-dwelling Morlocks? "As I recall," Goldman told me recently, "in that book it didn't work out very well for the Eloi."

When lifespans extend indefinitely, the effects are felt throughout the life cycle, but the biggest social impact may be on the young. According to Joshua Goldstein, a demographer at Princeton, adolescence will in the future evolve into a period of experimentation and education that will last from the teenage years into the midthirties. In a kind of *wanderjahr* prolonged for decades, young people will try out jobs on a temporary basis, float in and out of their parents' homes, hit the Europass-and-hostel circuit, pick up extra courses and degrees, and live with different people in different places. In the past the transition from youth to adulthood usually followed an orderly sequence: education, entry into the labor force, marriage, and parenthood. For tomorrow's thirtysomethings, suspended in what Goldstein calls "quasi-adulthood," these steps may occur in any order.

From our short-life-expectancy point of view, quasi-adulthood may seem like a period of socially mandated fecklessness — what Leon Kass, the chair of the President's Council on Bioethics, has decried as the coming culture of "protracted youthfulness, hedonism, and sexual license." In Japan, ever in the demographic forefront, as many as one out of three young adults is either unemployed or working part-time, and many are living rent-free with their parents. Masahiro Yamada, a sociologist at Tokyo Gakugei University, has sarcastically dubbed them *parasaito shinguru,* or "parasite singles." Adult offspring who live with their parents are common in aging Europe, too. In 2003 a report from the British Prudential financial-services group awarded the 6.8 million British in this category the mocking name of "kippers" — "kids in parents' pockets eroding retirement savings."

To Kass, the main cause of this stasis is "the successful pursuit of longer life and better health." Kass's fulminations easily lend themselves to ridicule. Nonetheless, he is in many ways correct. According to Yuji Genda, an economist at Tokyo University, the drifty lives of parasite singles are indeed a byproduct of increased longevity, mainly because longer-lived seniors are holding on to their jobs. Japan, with the world's oldest population, has the highest percentage of working senior citizens of any developed nation: one out of three men over sixty-five is still on the job. Everyone in the nation, Genda says, is "tacitly aware" that the old are "blocking the door."

In a world of two-hundred-year-olds "the rate of rise in income and status perhaps for the first hundred years of life will be almost negligible," the crusty maverick economist Kenneth Boulding argued in a prescient article from 1965. "It is the propensity of the old, rich, and powerful to die that gives the young, poor, and powerless hope." (Boulding died in 1993, opening up a position for another crusty maverick economist.)

Kass believes that "human beings, once they have attained the burdensome knowledge of good and bad, should not have access to the tree of life." Accordingly, he has proposed a straightforward way to prevent the problems of youth in a society dominated by the old: "Resist the siren song of the conquest of aging and death." Senior citizens, in other words, should let nature take its course once humankind's biblical seventy-year lifespan is up. Unfortunately, this solution is self-canceling, since everyone who agrees with it is eventually eliminated. Opponents, meanwhile, live on and on. Kass, who is sixty-six, has another four years to make his case.

Increased longevity may add to marital strains. The historian Lawrence Stone was among the first to note that divorce was rare in previous centuries partly because people died so young that bad unions were often dissolved by early funerals. As people lived longer, Stone argued, divorce became "a functional substitute for death." Indeed, marriages dissolved at about the same rate in 1860 as in 1960, except that in the nineteenth century the dissolution was more often due to the death of a partner, and in the twentieth century to divorce. The corollary that children were as likely to live in households without both biological parents in 1860 as in 1960 is also true. Longer lifespans are far from the only reason for today's higher divorce rates, but the evidence seems clear that they play a role. The prospect of spending another twenty years sitting across the breakfast table from a spouse whose charm has faded must have already driven millions to divorce lawyers. Adding an extra decade or two can only exacerbate the strain.

Worse, child-rearing, a primary marital activity, will be even more difficult than it is now. For the past three decades, according to Ben J. Wattenberg, a senior fellow at the American Enterprise Institute, birth rates around the world have fallen sharply as women have taken advantage of increased opportunities for education and work outside the home. "More education, more work, lower fertility," he says. The title of Wattenberg's latest book, published in October 2004, sums up his view of tomorrow's demographic prospects: *Fewer.* In his analysis, women's continuing movement outside the home will lead to a devastating population crash — the mirror image of the population boom that shaped so much of the past century. Increased longevity will only add to the downward pressure on birth rates, by making childbearing even more difficult. During their twenties, as Goldstein's quasi-adults, men and women will be unmarried and relatively poor. In their thirties and forties they will finally grow old enough to begin meaningful careers — the worst time to have children. Waiting still longer will mean entering the maelstrom of reproductive technology, which seems likely to remain expensive, alienating, and prone to complications. Thus the parental paradox: increased longevity means *less* time for pregnancy and child-rearing, not more.

Even when women manage to fit pregnancy into their careers, they will spend a smaller fraction of their lives raising children than ever before. In the mid-nineteenth century, white women in the United States had a life expectancy of about forty years and typically bore five or six children. (I specify Caucasians because records were not kept for African Americans.) These women literally spent more than half their lives caring for offspring. Today U.S. white women have a life expectancy of nearly eighty and bear an average of 1.9 children — below replacement level. If a woman spaces two births close together, she may spend only a quarter of her days in the company of offspring under the age of eighteen. Children will become ever briefer parentheses in long, crowded adult existences. It seems inevitable that the bonds between generations will fray.

Purely from a financial standpoint, parenthood has always been a terrible deal. Mom and Dad fed, clothed, housed, and educated the kids but received little in the way of tangible return. Ever since humankind began acquiring property, wealth has flowed from older generations to younger ones. Even in those societies where children herded cattle and tilled the land for their aged progenitors, the older generation consumed so little and died off so quickly that the net movement of assets and services was always downward.

"Of all the misconceptions that should be banished from discussions of aging," F. Landis MacKellar, an economist at the International Institute for Applied Systems Analysis, in Austria, wrote in the journal *Population and Development Review* in 2001, "the most persistent and egregious is that in some simpler and more virtuous age children supported their parents."

This ancient pattern changed at the beginning of the twentieth century, when government pension and social security schemes spread across Europe and into the Americas. Within the family parents still gave much more than they received, according to MacKellar, but under the new state plans the children in effect banded together outside the family and collectively reimbursed the parents. In the United States workers pay less to Social Security than they eventually receive; retirees are subsidized by the contributions of younger workers. But on the broadest level financial support from the young is still offset by the movement of assets within families — a point rarely noted by critics of "greedy geezers."

Increased longevity will break up this relatively equitable arrangement. Here concerns focus less on the super-rich than on middle-class senior citizens, those who aren't surfing the crest of compound interest. These people will face a Hobson's choice. On the one hand, they will be unable to retire at sixty-five, because the young would end up bankrupting themselves to support them — a reason why many would-be reformers propose raising the retirement age. On the other hand, it will not be feasible for most of tomorrow's nonagenarians and centenarians to stay at their desks, no matter how fit and healthy they are.

The case against early retirement is well known. In economic jargon the ratio of retirees to workers is known as the "dependency ratio," because through pension and Social Security payments people who are now in the work force funnel money to people who have left it. A widely cited analysis by three economists at the Organization for Economic Cooperation and Development estimated that in 2000 the overall dependency ratio in the United States was 21.7 retirees for every 100 workers, meaning (roughly speaking) that everyone older than sixty-five had five younger workers contributing to his pension. By 2050 the dependency ratio will have almost doubled, to 38 per 100; that is, each retiree will be supported by slightly more than two current workers. If old-age benefits stay the same, in other words, the burden on younger workers, usually in the form of taxes, will more than double.

This may be an underestimate. The OECD analysis did not assume any dramatic increase in longevity or the creation of any entitlement program to pay for longevity care. If both occur, as gerontological optimists predict, the number of old will skyrocket, as will the cost of maintaining them. To adjust to these "very bad fiscal effects," says the OECD economist Pablo Antolin, one of the report's coauthors, societies have only two choices: "raising the retirement age or cutting the benefits." He continues, "This is arithmetic — it can't be avoided." The recent passage of a huge new prescription-drug program by an administration and Congress dominated by the "party of small government" suggests that benefits will not be cut. Raising the age of retirement might be more feasible politically, but it would lead to a host of new problems — see today's Japan.

In the classic job pattern, salaries rise steadily with seniority. Companies underpay younger workers and overpay older workers as a means of rewarding employees who stay at their jobs. But as people have become more likely to shift firms and careers, the pay increases have become powerful disincentives for companies to retain employees in their fifties and sixties. Employers already worried about the affordability of older workers are not likely to welcome calls to raise the retirement age; the last thing they need is to keep middle managers around for another twenty or thirty years. "There will presumably be an elite group of super-rich who would be immune to all these pressures," Ronald Lee, an economic demographer at the University of California at Berkeley, says. "Nobody will kick Bill Gates out of Microsoft as long as he owns it. But there will be a lot of pressure on the average old person to get out."

In Lee's view, the financial downsizing need not be inhumane. One model is the university, which shifted older professors to emeritus status, reducing their workload in exchange for reduced pay. Or, rather, the university *could* be a model: age-discrimination litigation and professors' unwillingness to give up their perks, Lee says, have largely torpedoed the system. "It's hard to reduce someone's salary when they are older," he says. "For the person, it's viewed as a kind of disgrace. As a culture we need to get rid of that idea."

*

The Pentagon has released few statistics about the hundreds or thousands of insurgents captured in Afghanistan and Iraq, but one can be almost certain that they are disproportionately young. Young people have ever been in the forefront of political movements of all stripes. University students protested Vietnam, took over the U.S. embassy in Tehran, filled Tiananmen Square, served as the political vanguard for the Taliban. "When we are forty," the young writer Filippo Marinetti promised in the 1909 *Futurist Manifesto,* "other younger and stronger men will probably throw us in the wastebasket like useless manuscripts — we want it to happen!"

The same holds true in business and science. Steve Jobs and Stephen Wozniak founded Apple in their twenties; Albert Einstein dreamed up special relativity at about the same age. For better and worse, young people in developed nations will have less chance to shake things up in tomorrow's world. Poorer countries, where the old have less access to longevity treatments, will provide more opportunity, political and financial. As a result, according to Fred C. Iklé, an analyst with the Center for Strategic and International Studies, "it is not fanciful to imagine a new cleavage opening up in the world order." On one side would be the "'bioengineered' nations," societies dominated by the "becalmed temperament" of old people. On the other side would be the legions of youth — "the protagonists," as the political theorist Samuel Huntington has described them, "of protest, instability, reform, and revolution."

Because poorer countries would be less likely to be dominated by a gerontocracy, tomorrow's divide between old and young would mirror the contemporary division between rich northern nations and their poorer southern neighbors. But the consequences might be different — unpredictably so. One assumes, for instance, that the dictators who hold sway in Africa and the Middle East would not hesitate to avail themselves of longevity treatments, even if few others in their societies could afford them. Autocratic figures like Arafat, Franco, Perón, and Stalin often leave the scene only when they die. If the human lifespan lengthens greatly, the dictator in Gabriel García Márquez's *The Autumn of the Patriarch,* who is "an indefinite age somewhere between 107 and 232 years," may no longer be regarded as a product of magical realism.

Bioengineered nations, top-heavy with the old, will need to replenish their labor forces. Here immigration is the economist's traditional solution. In abstract terms, the idea of importing young workers from poor regions of the world seems like a win-win solution: the young get jobs, the old get cheap service. In practice, though, host nations have found that the foreigners in their midst are stubbornly . . . foreign. European nations are wondering whether they really should have let in so many Muslims. In the United States, traditionally hospitable to migrants, bilingual education is under attack and the southern border is increasingly locked down. Japan, preoccupied by *Nihonjinron* (theories of "Japaneseness"), has always viewed immigrants with suspicion if not hostility. Facing potential demographic calamity, the Japanese government has spent millions trying to develop a novel substitute for immigrants: robots smart and deft enough to take care of the aged.

According to Ronald Lee, the Berkeley demographer, rises in life expectancy have in the past stimulated economic growth. Because they arose mainly from reductions in infant and child mortality, these rises produced more healthy young workers, which in turn led to more-productive societies. Believing they would live a long time, those young workers saved more for retirement than their forebears, increasing society's stock of capital — another engine of growth. But these positive effects are offset when increases in longevity come from old people's neglecting to die. Older workers are usually less productive than younger ones, earning less and consuming more. Worse, the soaring expenses of entitlement programs for the old are likely, Lee believes, "to squeeze out government expenditures on the next generation," such as education and childhood public-health programs. "I think there's evidence that something like this is already happening among the industrial countries," he says. The combination will force a slowdown in economic growth: the economic pie won't grow as fast. But there's a bright side, at least potentially. If the fall in birth rates is sufficiently vertiginous, the number of people sharing that relatively smaller pie may shrink fast enough to let everyone have a bigger piece. One effect of the longevity-induced "birth dearth" that Wattenburg fears, in other words, may be higher per capita incomes.

For the past thirty years the United States has financed its budget deficits by persuading foreigners to buy U.S. Treasury bonds. In the nature of things, most of these foreigners have lived in other wealthy nations, especially Japan and China. Unfortunately for the United States, those other countries are marching toward longevity crises of their own. They, too, will have fewer young, productive workers. They, too, will be paying for longevity treatments for the old. They, too, will be facing a grinding economic slowdown. For all these reasons they may be less willing to finance our government. If so, Uncle Sam will have to raise interest rates to attract investors, which will further depress growth — a vicious circle.

Longevity-induced slowdowns could make young nations more attractive as investment targets, especially for the cash-strapped pension-and-insurance plans in aging countries. The youthful and ambitious may well follow the money to where the action is. If Mexicans and Guatemalans have fewer rich old people blocking their paths, the river of migration may begin to flow in the other direction. In a reverse brain drain, the Chinese coast guard might discover half-starved American postgraduates stuffed into the holds of smugglers' ships. Highways out of Tijuana or Nogales might bear road signs telling drivers to watch out for *norteamericano* families running across the blacktop, the children's Hello Kitty backpacks silhouetted against a yellow warning background.

Given that today nobody knows precisely how to engineer major increases in the human lifespan, contemplating these issues may seem premature. Yet so many scientists believe that some of the new research will pay off, and that lifespans will stretch like taffy, that it would be shortsighted not to consider the consequences. And the potential changes are so enormous and hard to grasp that they can't be understood and planned for at the last minute. "By definition," says Aubrey de Grey, the Cambridge geneticist, "you live with longevity for a very long time."

CHRIS MOONEY

The Dover Monkey Trial

FROM *Seed*

RESIGNING from the Dover Area School Board was the last thing that Jeff and Carol "Casey" Brown thought they'd ever have to do. They had poured their lives into their community's educational problems, Jeff for five years and Casey for ten. But in a late-night discussion on October 16, 2004 — the night of their twentieth wedding anniversary, as it happened — they realized they had little choice but to quit. Dover, Pennsylvania, the tiny township of 1,800 where they'd made their home for the past twenty-two years, had been radically altered. Former friends and school board colleagues were now bitter enemies, divided over a simple matter — what should be taught in ninth-grade biology at Dover High School.

Casey Brown, a tall woman with an uncompromising bent and a parliamentarian's mastery of school board rules, precedents, and procedures, first ran for the board to advocate better treatment of students with learning disabilities, such as her daughter. Jeff ran for the board because he was fed up with what he viewed as rampant cronyism and corruption. "I made a little cardboard button, took off a day of work, and got elected," he remembers.

That was back when Dover was still a sleepy place, long before most Americans had heard of the town or the *New York Times* assigned a reporter to cover its school board races. And it was long before the Dover board drove away the Browns, made national headlines, triggered a lawsuit for refuting Charles Darwin's theory of evolution, and introduced students to the concept of intelligent design, or ID.

With the highly visible federal trial that began on September 27, amid a national and international uproar prompted by President Bush's own endorsement of ID, the press depicted Dover as the twenty-first-century equivalent of Dayton, Tennessee, site of the famous 1925 Scopes monkey trial. But there's a key difference between the Scopes era and today: antievolutionists seem to be abandoning the Dover confrontation like a sinking ship. They have plans elsewhere — particularly in the state of Kansas. As a result, Dover represents something very different and perhaps more poignant. It's among the first towns to fall victim to a divisive religious and scientific battle that is building to a fever pitch, one that promises to tear apart many more communities before it's finally settled.

Jeff and Casey were on the front lines as the fight engulfed their town; in fact, they inadvertently facilitated it by choosing the wrong political allies. After their election to the nine-person school board, the Browns began to seek like-minded acquaintances to run for seats alongside them. They turned to a group of conservative Christians, who were soon elected. Together, they formed a majority on the board. The Dover area was home to a wide diversity of sects, including Mennonites, Lutherans, Brethren, and Amish, and had a long-standing live-and-let-live tradition. "All of us were good friends, and religion didn't really enter into it," recalls Casey. Then things began to change.

In 2003 the Dover board proposed sending a letter to the U.S. Supreme Court that defended using the phrase "under God" in the Pledge of Allegiance. Jeff, himself a Christian and a former Sunday school teacher, balked, asking if the board would also support the phrase "under Allah." Casey, an Episcopalian, also declared her opposition. But their views were in the minority, and the letter was sent.

In 2004, as the Seattle-based Discovery Institute — the national hub of the intelligent-design movement — impelled a new wave of fights over the teaching of evolution across the country, the Dover board pushed its own pro-ID agenda. The curriculum committee chair, William Buckingham, a conservative Christian, denounced a widely accepted biology textbook as being "laced with Darwinism." The fifty-seven-year-old former police officer and prison guard (who often denounced the notion of the separation of church and state as a myth) encouraged the search for a book that supported creationism, stating, "Two thousand years ago, someone died on a cross. Can't someone take a stand for him?" In another candid (and legally liable) moment, he added, "This country wasn't founded on Muslim beliefs or evolution. This country was founded on Christianity, and our students should be taught as such." (Although these statements were reported in local papers, Buckingham later contested them.)

The Browns were shocked, but their former friend had strong support on the board and in the community. Shortly thereafter he announced a mysterious "donation" to the high school: fifty copies of a textbook titled *Of Pandas and People*, published by the Texas-based Foundation on Thought and Ethics, a Christian organization. The book argues that natural processes cannot sufficiently explain the complexity of life and that "intelligent causes" must be invoked instead. Officially, the donation was anonymous. Unofficially — as was later alleged in court filings — the books arrived after Buckingham solicited donations for them at his own church.

According to Jeff and Casey, Buckingham and his supporters had by then stopped listening to the couple's dissenting opinions. "It's one thing to be at war with your political enemies, and I'd done that for years," says Jeff. "But when my friends turned into enemies, I just reached the end of the rope."

On Monday, October 18, two days after the Browns' anniversary, the board sealed their fate, and its own. It voted six to three to endorse the following change to the biology curriculum: "Students will be made aware of gaps/problems in Darwin's theory and of other theories of evolution including, but not limited to, intelligent design."

The Browns promptly announced their resignations; Casey read an eloquent statement declaring that as the evolution fight unfolded, board members had twice demanded to know whether she had been born again. "It has become increasingly evident that, in the direction this board has now chosen to go, holding a certain religious belief is of paramount importance," she said. In a deposition, Jeff Brown would later charge that after he and Casey decided to resign, they were labeled "atheists" by one pro-ID board member, Alan Bonsell.

Shortly after the vote on ID, eleven parents of Dover High School students engaged the American Civil Liberties Union, Americans United for the Separation of Church and State, and the Pennsylvania law firm of Pepper Hamilton. They filed suit over the board's decision. "I want my daughter to have her religious education, but I want to be responsible for it, or maybe the church we attend," says plaintiff Steve Stough, a Republican and a Christian who, like Jeff and Casey Brown, accepts evolution. "And no matter how you describe it, this whole thing was a shot at religious education."

Dover's science and religion rift goes back several years, as illustrated by an event that took place over summer recess in 2002. While Dover high school students and teachers were on break, a high school janitor removed a large student-painted mural from its place in one of the school's science labs and set it on fire. The massive, colorful piece of artwork — taking up two four-by-eight-foot plywood sheets — depicted an evolutionary progression of ape-like humanoids running across grassland as they gradually became modern man. It had been commissioned by the science department itself, and it took its student creator a full semester to complete.

In picking a fight over evolution, the Dover board had exacerbated precisely what the authors of the U.S. Constitution sought to prevent with the First Amendment: Religious schisms in American communities. This fall the Dover Area School Board's actions will be judged according to the amendment — as well as by precedents set by the long string of evolution lawsuits that punctuated U.S. history during the twentieth century.

The fight against evolution in America has never really ended; it has only changed form — for legal and cultural reasons rather than scientific ones. The 1925 Scopes monkey trial — memorialized in the play and film *Inherit the Wind* — took place after John Scopes, a first-year science teacher, deliberately violated Tennessee's explicit antievolution statute, which stated that "it shall be unlawful for any teacher to teach any law that denies the story of the Divine Creation of man as taught in the Bible, and to teach instead that man has descended from a lower order of animals." Prosecutor William Jennings Bryan defended the literal authority of the Bible and argued that science and religion were locked in inevitable conflict — evolution and biblical truth could never be reconciled. Clarence Darrow, the ACLU criminal defense lawyer, sought to show that the theory of evolution was perfectly compatible with equally valid, but nonliteral, readings of Scripture.

What is less frequently remembered about the Scopes trial is its somewhat ignominious conclusion. Scopes was convicted for doing what he had inarguably done: violated existing law. Because Scopes was later acquitted on a technicality, the ACLU could not appeal and seek a strong precedent in support of the separation of religion and science education. The cultural legacy of this trial clearly advanced the creationist cause in America. Other states passed antievolution laws; publishers of high school biology textbooks self-censored to conform. Then in the 1960s, after the Soviets launched *Sputnik*, the U.S. government increased spending on scientific research and education, promoting evolution at the high school level. The challenge to creationist sentiment and school board policy ignited a second chapter in the history of battles over the teaching of evolution in schools. This time, though, the antievolutionists were on the defensive.

In 1968 the U.S. Supreme Court overturned an Arkansas antievolution law very similar to the Tennessee law violated by Scopes, calling it an affront to church-state separation and the First Amendment. In the wake of this new precedent, the antievolutionist legal strategy advocated "equal time" legislation, calling for the inclusion of both evolution and creationism (which creationists now labeled "scientific") in high school biology. In 1981 *McLean v. Arkansas* was fought over that state's "balanced treatment for creation-science and evolution-science" Act 590. In a case closely resembling what is now happening in Dover, the ACLU challenged the law's constitutionality, leading to a lengthy and involved federal trial in which both sides, wielding dueling "experts," claimed to have science on their side. The ACLU sought to prove that "creation science" was in fact nothing of the kind, and Judge William Overton ruled strongly in the group's favor.

In 1987 the U.S. Supreme Court ruled that Louisiana's "balanced treatment" law was also unconstitutional, favorably citing *McLean*. Both courts pronounced that "creation science" was, in essence, a fraud — religion masquerading as a valid scientific explanation just to get past legal barriers. With their "creation science" strategies struck down by the Supreme Court, antievolutionists almost immediately launched another tactic: they morphed into defenders of "intelligent design."

According to court filings, certain Dover board members pushed for the teaching of outright creationism long before they accepted a donation of antievolution textbooks from a local church. It was also long before they embraced ID — which is itself inherently theological, as the ACLU and colitigants argue, because it postulates a designer who is "intelligent."

"The purposes and effect of the [intelligent design] policy are to advance and endorse the specific religious viewpoint and beliefs encompassed by the assertion or argument of intelligent design," the Dover lawsuit charges. When set in the context of Supreme Court legal precedents, the Dover board's actions would appear to leave it thoroughly exposed — relatively easy pickings for civil-libertarian lawyers who have defended evolution against religious onslaughts again and again in recent history.

If battling over culturally divisive issues like evolution can be destructive to a community, it can also be quite expensive in terms of legal fees; Dover, which Casey Brown calls a "bedroom community," is a relatively poor area with a modest tax base. If, as seems increasingly likely, the Dover board — represented pro bono by the Thomas More Law Center of Ann Arbor, Michigan, which describes itself as dedicated to "defending and promoting the religious freedom of Christians" — loses its case, the fiscal penalty (comprising the considerable legal expenses for the ACLU and colitigants) could be substantial. The religious strife that Dover has already experienced may pale in comparison to the attacks and finger-pointing that will assuredly follow if Dover taxpayers have to underwrite a lawsuit that their elected school board inflicted upon the district.

From a practical standpoint, the emphasis on having to defend a federal lawsuit has almost certainly taken a toll on education in Dover. "How many hundreds and thousands of hours have been wasted on this already, by the administrators, the support staff, the teachers?" asks Casey Brown. "The time that was being spent on ID was not being spent on identifying at-risk kids," adds her husband.

And for what greater cause is Dover making all these sacrifices? Sadly, it may be for none at all. Even as the national media lionizes the Dover case as a new Scopes trial, top-tier antievolutionists have strategically backed away from it. In June, three leading Discovery Institute–affiliated ID advocates had their names removed from the list of experts slated to testify in court on behalf of the Dover school district. This, despite the fact that several Discovery fellows penned a 2000 *Utah Law Review* article, claiming that introducing ID into science classes — exactly the Dover board's strategy — is constitutionally permissible and legally defensible.

No longer the second coming of the Scopes trial, the Dover situation is looking more and more like the mess nobody in the creationist camp wants to clean up. Even the notorious Republican senator Rick Santorum, who last year stood by the Dover board's actions like a proud parent, is now pulling back from openly pushing for intelligent design in science classes. Santorum has relinquished his pro-ID stance; the Dover board, however, cannot. The national ID movement is backing away from Dover as though it's a village that must be sacrificed in the early stages of a long military campaign.

The Discovery Institute's updated strategy, which doesn't explicitly introduce ID into classes, is called "teach the controversy." Having manufactured a national debate through its widespread questioning of evolutionary biology — and through at least implicitly encouraging actions like that of Dover's school board — the Discovery Institute now points to the alleged "controversy" itself as a topic that American students need to understand. It's quite effective: school board members are only encouraging "critical

thinking" about evolution. ID proponents can watch as evolution is questioned in class; they won't need to run the legal risk of explicitly advocating religious views or talking about an "intelligent designer."

After the Dover case runs its course, or perhaps even sooner, some locality's "teach the controversy" policy will probably spark the next legal battle over evolution; Kansas seems the likely frontrunner. Although it hasn't finalized anything as of this writing, the antievolution majority on the Kansas State Board of Education has been pushing a Discovery Institute–friendly "critical analysis" of the teaching of evolution, intended for the state's science standards. They're also pushing a redefinition of science for the board's purposes, so that it would no longer be limited to naturally occurring phenomena — providing ID with the chance to call itself science. But Kansas, unlike Dover, has carefully avoided introducing the words "intelligent design" into the standards, a strategic posture that may put it on somewhat stronger legal footing.

That's no consolation for Dover, which won't easily heal from its battle. The nature of the Dover situation was apparent at the board's August meeting, where parents in the audience discussed the abrupt disappearance of William Buckingham, the political leader of the Dover board's actions.

After the media descended and the lawsuit was filed, Buckingham denied making his infamous statement about defending Jesus in America's classrooms. Not long thereafter, citing health problems, he sold his house and moved to North Carolina. At the meeting, the school superintendent read Buckingham's resignation statement. Though they'd followed his lead into potentially costly litigation, none of his former school board allies made use of the opportunity to comment on his influential tenure.

Jeff and Casey Brown were not present that day, though their side of the fight was well represented: sitting in the back of the room were the parents who had filed suit against the board. "For the person who spearheaded this whole movement, to see him bail at this point in time disturbs me," said Steve Stough, one of the plaintiffs.

Buckingham's southward flight seems symbolic of the Dover board's problematic legal situation. While there is a new religious challenge to evolution afoot, Dover seems increasingly unlikely to represent the deciding battle — yet the town's experience is all the more revealing for precisely that reason. Rather than fighting science with science, Dover illustrates that the evolution conflict on the horizon is likely to feature ever-shifting legal strategies, broken allegiances, and communities hung out to dry. It will be a conflict, in short, in which the very essence of the antievolutionists' tactics demonstrates the legal and scientific weakness of their position.

DENNIS OVERBYE

Remembrance of Things Future

FROM *The New York Times*

THERE WAS A CONFERENCE for time travelers at MIT earlier this spring. I'm still hoping to attend, and although the odds are slim, they are apparently not zero, despite the efforts and hopes of deterministically minded physicists who would like to eliminate the possibility of your creating a paradox by going back in time and killing your grandfather.

"No law of physics that we know of prohibits time travel," said J. Richard Gott, a Princeton astrophysicist. Gott, the author of the 2001 book *Time Travel in Einstein's Universe: The Physical Possibilities of Travel Through Time,* is one of a small breed of physicists who spend part of their time (and their research grants) thinking about wormholes in space, warp drives, and other cosmic constructions that "absurdly advanced civilizations" might use to travel through time.

It's not that physicists expect to be able to go back and attend Woodstock, drop by the Bern patent office to take Einstein to lunch, see the dinosaurs, or investigate John F. Kennedy's assassination.

In fact, they're pretty sure those are absurd dreams and are all bemused by the fact that they can't say why. They hope such extreme theorizing could reveal new features, gaps, or perhaps paradoxes or contradictions in the foundations of Physics As We Know It and point the way to new ideas. "Traversable wormholes are primarily useful as a 'gedanken experiment' to explore the limitations of general relativity," said Francisco Lobo of the University of Lisbon.

If general relativity, Einstein's theory of gravity and spacetime, allows for the ability to go back in time and kill your grandfather, asks David Z. Albert, a physicist and philosopher at Columbia University, "how can it be a logically consistent theory?" In his recent book *The Universe in a Nutshell,* Stephen W. Hawking wrote, "Even if it turns out that time travel is impossible, it is important that we understand why it is impossible."

When it comes to the nature of time, physicists for the most part are at as much of a loss as the rest of us who seem hopelessly swept along in its current. The mystery of time is connected with some of the thorniest questions in physics, as well as in philosophy, like why we remember the past but not the future, how causality works, why you can't stir cream out of your coffee or put perfume back in a bottle.

But some theorists think that has to change. Just as Einstein needed to come up with a new concept of time in order to invent relativity one hundred years ago this year, so physicists say that a new insight into time — or beyond it — may be required to crack profound problems like how the universe began, what happens at the center of a black hole, or how to marry relativity and quantum theory into a unified theory of nature.

Space and time, some quantum gravity theorists say, are most likely a sort of illusion — or, less sensationally, an "approximation" — doomed to be replaced by some more fundamental idea. If only they could think of what that idea is. "By convention there is space, by convention time," David J. Gross, director of the Kavli Institute for Theoretical Physics and a winner of last year's Nobel Prize, said recently, paraphrasing the Greek philosopher Democritus. "In reality there is . . . ?" His voice trailed off.

The issues raised by time travel are connected to these questions, said Lawrence Krauss, a physicist at Case Western Reserve University in Cleveland and the author of the book *The Physics of Star Trek.* "The minute you have time travel you have paradoxes," he said, explaining that if you can go backward in time you confront fundamental issues like cause and effect or the meaning of your own identity if there can be two of you at once. A refined theory of time would have to explain "how a sensible world could result from something so nonsensical. That's why time travel is philosophically important and has captivated the public, who care about these paradoxes."

At stake, said philosopher David Albert, the author of his own time book, *Time and Chance,* is "what kind of view science presents us of the world. Physics gets time wrong, and time is the most familiar thing there is."

We all feel time passing in our bones, but ever since Galileo and Newton in the seventeenth century began using time as a coordinate to help chart the motion of cannonballs, time — for physicists — has simply been an "addendum in the address of an event," Albert said. "There is a feeling in philosophy that this picture leaves no room for locutions about flow and the passage of time we experience."

Then there is what physicists call "the arrow of time" problem. The fundamental laws of physics don't care what direction time goes, Albert pointed out. Run a movie of billiard balls colliding or planets swirling around in their orbits in reverse and nothing will look weird, but if you run a movie of a baseball game in reverse, people will laugh.

Einstein once termed the distinction between past, present, and future "a stubborn illusion," but as David Albert said, "It's hard to imagine something more basic than the distinction between the future and the past."

The Birth of an Illusion

Space and time, the philosopher Augustine famously argued 1,700 years ago, are creatures of existence and the universe, born with it, not separately standing features of eternity. That is the same answer Einstein came up with in 1915 when he finished his general theory of relativity. That theory explains how matter and energy warp the geometry of space and time to produce the effect we call gravity. It also predicted, somewhat to Einstein's dismay, the expansion of the universe, which forms the basis of modern cosmology.

But Einstein's theory is incompatible, mathematically and philosophically, with the quirky rules known as quantum mechanics, which describe the microscopic randomness that fills this elegantly curved, expanding spacetime. According to relativity, nature is continuous, smooth, and orderly; in quantum theory the world is jumpy and discontinuous. The sacred laws of physics are correct only on average. Until the pair are married in a theory of so-called quantum gravity, physics has no way to investigate what happens in the big bang, when the entire universe is so small that quantum rules apply.

If looked at closely enough, with an imaginary microscope that could see lengths down to 10^{-33} centimeter, quantum gravity theorists say, even ordinary space and time dissolve into a boiling mess that John Wheeler, the Princeton physicist and phrasemaker, called "spacetime foam." At that level of reality, which exists underneath all our fingernails, clocks and rulers as we know them cease to exist. "Everything we know about stops at the big bang, the big crunch," said Raphael Bousso, a physicist at the University of California, Berkeley.

What happens to time at this level of reality is anybody's guess. Lee Smolin, of the Perimeter Institute for Theoretical Physics in Waterloo, Ontario, said, "There are several different, very different, ideas about time in quantum gravity." One view, he explained, is that space and time "emerge" from this foamy substrate when it is viewed at larger scales. Another is that space emerges but that time or some deeper relations of cause and effect are fundamental.

Fotini Markopoulou Kalamara of the Perimeter Institute described time as, if not an illusion, an approximation, "a bit like the way you can see the river flow in a smooth way even though the individual water molecules follow much more complicated patterns." She added in an e-mail message: "I have always thought that there has to be some basic fundamental notion of causality, even if it doesn't look at all like the one of the spacetime we live in. I can't see how to get causality from something that has none; neither have I ever seen anyone succeed in doing so."

Physicists say they have a sense of how space can emerge, because of recent advances in string theory, the putative theory of everything, which posits that nature is composed of wriggling little strings.

Calculations by Juan Maldacena of the Institute for Advanced Study in Princeton and by others have shown how an extra dimension of space can pop mathematically into being almost like magic, the way the illusion of three dimensions can appear in the holograms on bank cards. But string theorists admit that they don't know how to do the same thing for time yet.

"Time is really difficult," said Cumrun Vafa, a Harvard string theorist. "We have not made much progress on the emergence of time. Once we make progress we will make progress on the early universe, on high-energy physics and black holes. We are out on a limb trying to understand what's going on here."

Bousso, an expert on holographic theories of spacetime, said that in general relativity time gets no special treatment. He said he expected both time and space to break down, adding, "We really just don't know what's going to go."

"There is a lot of mysticism about time," Bousso said. "Time is what a clock measures. What a clock measures is more interesting than you thought."

A Brief History of Time Travel

"If we could go faster than light, we could telegraph into the past," Einstein once said. According to the theory of special relativity — which he proposed in 1905 and which ushered $E = mc^2$ into the world and set the speed of light as the cosmic speed limit — such telegraphy is not possible, and there is no way of getting back to the past.

But somewhat to Einstein's surprise, in general relativity it is possible to beat a light beam across space. That theory, which Einstein finished in 1916, said that gravity resulted from the warping of spacetime geometry by matter and energy, the way a bowling ball makes a trampoline sag. And all this warping and sagging can create shortcuts through spacetime.

In 1949 Kurt Gödel, the Austrian logician and mathematician then at the Institute for Advanced Study, showed that in a rotating universe, according to general relativity, there were paths, technically called "closed timelike curves," you could follow to get back to the past. But it has turned out that the universe does not rotate very much, if at all.

Most scientists, including Einstein, resisted the idea of time travel until 1988, when Kip Thorne, a gravitational theorist at the California Institute of Technology, and two of his graduate students, Mike Morris and Ulvi Yurtsever, published a pair of papers concluding that the laws of physics may allow you to use wormholes, which are like tunnels through space connecting distant points, to travel in time. These holes, technically called Einstein-Rosen bridges, have long been predicted as a solution of Einstein's equations. But physicists dismissed them because calculations predicted that gravity would slam them shut.

Thorne was inspired by his friend the late Cornell scientist and author Carl Sagan, who was writing the science-fiction novel *Contact* (later made into a Jodie Foster movie) and was looking for a way to send his heroine, Eleanor Arroway, across the galaxy. Thorne and his colleagues imagined that such holes could be kept from collapsing and thus maintained to be used as a galactic subway, at least in principle, by threading them with something called Casimir energy (after the Dutch physicist Hendrik Casimir), which is a sort of quantum suction produced when two parallel metal plates are placed very close together. According to Einstein's equations, this suction, or negative pressure, would have an antigravitational effect, keeping the walls of the wormhole apart.

If one mouth of a wormhole was then grabbed by a spaceship and taken on a high-speed trip, according to relativity, its clock would run slow compared with the other end of the wormhole. So the wormhole would become a portal between two different times as well as places.

Thorne later said he had been afraid that the words "time travel" in the second paper's title would create a sensation and tarnish his students' careers, and he had forbidden Caltech to publicize it. In fact, their paper made time travel safe for serious scientists, and other theorists, including Frank Tipler of Tulane University and Stephen Hawking, jumped in. In 1991, for example, Richard Gott showed how another shortcut through spacetime could be manufactured using pairs of cosmic strings — dense tubes of primordial energy not to be confused with the strings of string theory, left over by the big bang in some theories of cosmic evolution — rushing past each other and warping space around them.

Harnessing the Dark Side

These speculations have been bolstered (not that time-machine architects lack imagination) with the unsettling discovery that the universe may be full of exactly the kind of antigravity stuff needed to grow and prop open a wormhole. Some mysterious "dark energy," astronomers say, is pushing space apart and accelerating the expansion of the universe. The race is on to measure this energy precisely and find out what it is.

Among the weirder and more disturbing explanations for this cosmic riddle is something called phantom energy, which is so virulently antigravitational that it would eventually rip apart planets, people, and even atoms, ending everything. As it happens, this bizarre stuff would also be perfect for propping open a wormhole, Francisco Lobo recently pointed out. "This certainly is an interesting prospect for an absurdly advanced civilization, as phantom energy probably comprises 70 percent of the universe," he wrote in an e-mail message. Sergey Sushkov of Kazan State Pedagogical University in Russia has made the same suggestion.

In a paper posted on the physics Web site arxiv.org/abs/gr-qc/0502099, Lobo suggested that as the universe was stretched and stretched under phantom energy, microscopic holes in the quantum "spacetime foam" might grow to macroscopic usable size. "One could also imagine an advanced civilization mining the cosmic fluid for the phantom energy necessary to construct and sustain a traversable wormhole," he wrote. Such a wormhole, he even speculated, could be used to escape the "big rip" in which a phantom-energy universe will eventually end.

But nobody knows if phantom, or exotic, energy is really allowed in nature, and most physicists would be happy if it is not. Its existence would lead to paradoxes, like negative kinetic energy, which could cause something to lose energy by speeding up, violating what is left of common sense in modern physics. Lawrence Krauss said, "From the point of view of realistic theories, phantom energy just doesn't exist."

But such exotic stuff is not required for all time machines — Gott's cosmic strings, for example. In another recent paper, Amos Ori of the Technion-Israel Institute of Technology in Haifa describes a time machine that he claims can be built by moving around colossal masses to warp the space inside a doughnut of regular empty space into a particular configuration, something an advanced civilization may be able to do in one hundred or two hundred years. The space inside the doughnut, he said, will then naturally evolve according to Einstein's laws into a time machine.

Ori admits that he doesn't know if his machine would be stable. Time machines could blow up as soon as you turned them on, say some physicists, including Hawking, who has proposed what he calls the "chronology protection" conjecture to keep the past safe for historians. Random microscopic fluctuations in matter and energy and space itself, they argue, would be amplified by going around and around the boundaries of the machine or the wormhole and finally blow it up.

Gott and his colleague Li-Xin Li have shown that there are at least some cases in which the time machine does not blow up. But until gravity marries quantum theory, they admit, nobody knows how to predict exactly what the fluctuations would be. "That's why we really need to know about quantum gravity," Gott said. "That's one reason people are interested in time travel."

Saving Grandpa

But what about killing your grandfather? In a well-ordered universe, that would be a paradox and shouldn't be able to happen, everybody agrees.

That was the challenge Joe Polchinski, now at the Kavli Institute for Theoretical Physics in Santa Barbara, California, issued to Thorne and his colleagues after their paper was published. Being a good physicist, Polchinski phrased the problem in terms of billiard balls. A billiard ball, he suggested, could roll into one end of a time machine, come back out the other end a little earlier, and collide with its earlier self, thereby preventing itself from entering the time machine to begin with.

Thorne and two students, Fernando Echeverria and Gunnar Klinkhammer, concluded after months of mathematical struggle that there was a logically consistent solution to the billiard matricide that Polchinski had set up. The ball would come back out of the time machine and deliver only a glancing blow to itself, altering its path just enough so that it would still hit the time machine. When it came back out, it would be aimed just so as to deflect itself rather than hitting full on. And so it would go, like a movie with a circular plot.

In other words, it's not a paradox if you go back in time and save your grandfather. And, added Polchinski, "It's not a paradox if you try to shoot your grandfather and miss."

"The conclusion is somewhat satisfying," Thorne wrote in his book *Black Holes and Time Warps: Einstein's Outrageous Legacy.* "It suggests that the laws of physics might accommodate themselves to time machines fairly nicely."

Polchinski agreed. "I was making the point that the grandfather paradox had nothing to do with free will, and they found a nifty resolution," he said in an e-mail message, adding, nevertheless, that his intuition still tells him time machines would lead to paradoxes.

Raphael Bousso said, "Most of us would consider it quite satisfactory if the laws of quantum gravity forbade time travel."

PAUL RAFFAELE

Out of Time

FROM *Smithsonian*

DEEP IN THE AMAZON JUNGLE, I stumble along a sodden track carved through steamy undergrowth, frequently sinking to my knees in the mud. Leading the way is a bushy-bearded, fiery-eyed Brazilian, Sydney Possuelo, South America's leading expert on remote Indian tribes and the last of the continent's great explorers. Our destination: the village of a fierce tribe not far removed from the Stone Age.

We're in the Javari Valley, one of the Amazon's "exclusion zones" — huge tracts of virgin jungle set aside over the past decade by the government of Brazil for indigenous Indians and off-limits to outsiders. Hundreds of people from a handful of tribes live in the valley amid misty swamps, twisting rivers, and sweltering rain forests bristling with anacondas, caimans, and jaguars. They have little or no knowledge of the outside world, and often face off against each other in violent warfare.

About half a mile in from the riverbank where we docked our boat, Possuelo cups his hands and shouts a melodious "Eh-heh." "We're near the village," he explains, "and only enemies come in silence." Through the trees, a faint "Eh-heh" returns his call.

We keep walking, and soon the sunlight stabbing through the trees signals a clearing. At the top of a slope stand about twenty naked Indians — the women with their bodies painted blood red, the men gripping formidable-looking clubs. "There they are," Possuelo murmurs, using the name they're called by other local Indians, "Korubo!" The group call themselves Dslala, but it's their Portuguese name I'm thinking of now: *caceteiros,* or "head bashers." I remember his warning of a half-hour earlier as we trudged through the muck: "Be on your guard at all times when we're with them, because they're unpredictable and very violent. They brutally murdered three white men just two years ago."

My journey several thousand years back in time began at the frontier town of Tabatinga, about 2,200 miles northwest of Rio de Janeiro, where a tangle of islands and sloping mud banks shaped by the mighty Amazon forms the borders of Brazil, Peru, and Colombia. There Possuelo and I boarded his speedboat, and he gunned it up the Javari River, an Amazon tributary. "Bandits lurk along the river, and they'll shoot to kill if they think we're worth robbing," he said. "If you hear gunfire, duck."

A youthful, energetic sixty-four, Possuelo is the head of the Department for Isolated Indians in FUNAI, Brazil's National Indian Bureau. He lives in the capital city, Brasília, but he's happiest when he's at his base camp just inside the Javari Valley exclusion zone, from which he fans out to visit his beloved Indians. It's the culmination of a dream that began as a teenager, when like many kids his age, he fantasized about living a life of adventure.

The dream began to come true forty-two years ago, when Possuelo became a *sertanista,* or "backlands expert" — drawn, he says, "by my wish to lead expeditions to remote Indians." A dying breed today, the sertanistas are peculiar to Brazil, Indian trackers charged by the government with finding tribes in hard-to-reach interior lands. Most sertanistas count themselves lucky to have made "first contact" — a successful initial nonviolent encounter between a tribe and the outside world — with one or two Indian tribes, but Possuelo has made first contact with no fewer than seven. He's also identified twenty-two sites where uncontacted Indians live, apparently still unaware of the larger world around them, except for the rare skirmish with a Brazilian logger or fisherman who sneaks into their sanctuary. At least four of these uncontacted tribes are in the Javari Valley. "I've spent months at a time in the jungle on expeditions to make first contact with a tribe, and I've been attacked many, many times," he says. "Colleagues have fallen at my feet, pierced by Indian arrows." Since the 1970s, in fact, 120 FUNAI workers have been killed in the Amazon jungles.

Now we're on the way to visit a Korubo clan he first made contact with in 1996. For Possuelo it's one of his regular check-in visits, to see how they're faring; for me it's a chance to be one of the few journalists ever to spend several days with this group of people, who know nothing about bricks or electricity or roads or violins or penicillin or Cervantes or tap water or China or almost anything else you can think of.

Our boat passes a river town named Benjamin Constant, dominated by a cathedral and a timber mill. Possuelo glares at both. "The church and loggers are my biggest enemies," he tells me. "The church wants to convert the Indians to Christianity, destroying their traditional ways of life, and the loggers want to cut down their trees, ruining their forests. It's my destiny to protect them."

At the time the portuguese explorer Pedro Cabral strode ashore in 1500 AD to claim Brazil's coast and vast inland for his king, perhaps as many as ten million Indians lived in the rain forests and deltas of the world's second-longest river. During the following centuries, sertanistas led white settlers into the wilderness to seize Indian lands and enslave and kill countless tribespeople. Hundreds of tribes were wiped out as rubber tappers, gold miners, loggers, cattle ranchers, and fishermen swarmed over the pristine jungles. And millions of Indians died from strange new diseases, like the flu and measles, for which they had no immunity.

When he first became a sertanista, Possuelo himself was seduced by the thrill of the dangerous chase, leading hundreds of search parties into Indian territory — no longer to kill the natives but to bring them out of their traditional ways and into Western civilization (while opening up their lands, of course, to outside ownership). By the early 1980s, though, he had concluded that the clash of cultures was destroying the tribes. Like Australia's Aborigines and Alaska's Inuit, the Indians of the Amazon Basin were drawn to the fringes of the towns that sprang up in their territory, where they fell prey to alcoholism, disease, prostitution, and the destruction of their cultural identity. Now only an estimated 350,000 Amazon Indians remain, more than half in or near towns. "They've largely lost their tribal ways," Possuelo says. The cultural survival of isolated tribes like the Korubo, he adds, depends on "our protecting them from the outside world."

In 1986 Possuelo created the Department for Isolated Indians and — in an about-face from his previous work — championed, against fierce opposition, a policy of discouraging contact with remote Indians. Eleven years later he defied powerful politicians and forced all non-Indians to leave the Javari Valley, effectively quarantining the tribes that remained. "I expelled the loggers and fishermen who were killing the Indians," he boasts.

Most of the outsiders were from Atalaia — at fifty miles downriver, the nearest town to the exclusion zone. As we pass the town, where a marketplace and huts spill down the riverbank, Possuelo tells a story. "Three years ago, more than three hundred men armed with guns and Molotov cocktails" — angry at being denied access to the valley's plentiful timber and bountiful fishing — "came up to the valley from Atalaia planning to attack my base," he says. He radioed the federal police, who quickly arrived in helicopters, and after an uneasy standoff, the raiders turned back. And now? "They'd still like to destroy the base, and they've threatened to kill me."

For decades, violent clashes have punctuated the long-running frontier war between the isolated Indian tribes and "whites" — the name that Brazilian Indians and non-Indians alike use to describe non-Indians, even though in multiracial Brazil many of them are black or of mixed race — seeking to profit from the rain forests. More than forty whites have been massacred in the Javari Valley, and whites have shot dead hundreds of Indians over the past century.

But Possuelo has been a target of settlers' wrath only since the late 1990s, when he led a successful campaign to double the size of the exclusion zones; the restricted territories now take up 11 percent of Brazil's huge landmass. That has drawn the attention of businessmen who wouldn't normally care much about whether a bunch of Indians ever leaves the forest, because in an effort to shield the Indians from life in the modern age, Possuelo has also safeguarded a massive slab of the earth's species-rich rain forests. "We've ensured that millions of hectares of virgin jungle are shielded from the developers," he says, smiling. And not everyone is as happy about that as he is.

About four hours into our journey from Tabatinga, Possuelo turns the speedboat into the mouth of the coffee-hued Itacuai River and follows that to the Itui River. We reach the entrance to the Javari Valley's Indian zone soon afterward. Large signs on the riverbank announce that outsiders are prohibited from venturing farther.

A Brazilian flag flies over Possuelo's base, a wooden bungalow perched on poles overlooking the river and a pontoon containing a medical post. We're greeted by a nurse, Maria da Graca Nobre, nicknamed Magna, and two fearsome-looking, tattooed Matis Indians, Jumi and Jemi, who work as trackers and guards for Possuelo's expeditions. Because the Matis speak a language similar to the lilting, high-pitched Korubo tongue, Jumi and Jemi will also act as our interpreters.

In his spartan bedroom, Possuelo swiftly exchanges his bureaucrat's uniform — crisp slacks, shoes, and a black shirt bearing a FUNAI logo — for his jungle gear: bare feet, ragged shorts, and a torn, unbuttoned khaki shirt. In a final flourish, he flings on a necklace hung with a bullet-size cylinder of antimalarial medicine, a reminder that he's had thirty-nine bouts with the disease.

The next day we head up the Itui in an outboard-rigged canoe for the land of the Korubo. Caimans doze on the banks while rainbow-hued parrots fly overhead. After half an hour, a pair of dugouts on the riverbank tell us the Korubo are near, and we disembark to begin our trek along the muddy jungle track.

When at last we come face to face with the Korubo in the sun-dappled clearing, about the size of two football fields and scattered with fallen trees, Jumi and Jemi grasp their rifles, warily watching the men with their war clubs. The Korubo stand outside a *maloca,* a communal straw hut built on a tall framework of poles, about twenty feet wide, fifteen feet high, and thirty feet long.

The seminomadic clan moves between four or five widely dispersed huts as their maize and manioc crops come into season, and it took Possuelo four lengthy expeditions over several months to catch up to them the first time. "I wanted to leave them alone," he says, "but loggers and fishermen had located them and were trying to wipe them out. So I stepped in to protect them."

They weren't particularly grateful. Ten months later, after intermittent contact with Possuelo and other FUNAI fieldworkers, the clan's most powerful warrior, Ta'van, killed an experienced FUNAI sertanista, Possuelo's close friend Raimundo Batista Magalhaes, crushing his skull with a war club. The clan fled into the jungle, returning to the maloca only after several months.

Now Possuelo points out Ta'van — taller than the others, with a wolfish face and glowering eyes. Ta'van never relaxes his grip on his sturdy war club, which is longer than he is and stained red. When I lock eyes with him, he glares back defiantly. Turning to Possuelo, I ask how it feels to come face to face with his friend's killer. He shrugs. "We whites have been killing them for decades," he says. Of course, it's not the first time that Possuelo has seen Ta'van since Magalhaes's death. But only recently has Ta'van offered a reason for the killing, saying simply, "We didn't know you then."

While the men wield the clubs, Possuelo says that "the women are often stronger," so it doesn't surprise me to see that the person who seems to direct the Korubo goings-on is a woman in her midforties, named Maya. She has a matronly face and speaks in a girlish voice, but hard dark eyes suggest an unyielding nature. "Maya," Possuelo tells me, smiling, "makes all the decisions." By her side is Washman, her eldest daughter, grim-faced and in her early twenties. Washman has "the same bossy manner as Maya," Possuelo adds with another smile.

Their bossiness may extend to ordering murders. Two years ago three warriors led by Ta'van and armed with clubs — other Indian tribes in the Javari Valley use bows and arrows in war, but the Korubo use clubs — paddled their dugout down the river until they came upon three white men just beyond the exclusion zone, cutting down trees. The warriors smashed the whites' heads to pulp and gutted them. Possuelo, who was in Atalaia when the attack occurred, rushed upriver to where the mutilated bodies lay, finding the murdered men's canoe "full of blood and pieces of skull."

As grisly as the scene was, Possuelo was not displeased when news of the killing spread quickly in Atalaia and other riverside settlements. "I prefer them to be violent," he says, "because it frightens off intruders." Ta'van and the others have not been charged, a decision Possuelo supports: the isolated Indians from the Javari Valley, he says, "have no knowledge of our law and so can't be prosecuted for any crime."

After Possuelo speaks quietly with Maya and the others for half an hour in the clearing, she invites him into the maloca. Jemi, Magna, and most of the clan follow, leaving me outside with Jumi and a pair of children, naked like their parents, who exchange shy smiles with me. A young spider monkey, a family pet, clings to one little girl's neck. Maya's youngest child, Manis, sits beside me, cradling a baby sloth, also a pet.

Even with Jumi nearby, I glance about warily, not trusting the head bashers. About an hour later, Possuelo emerges from the maloca. At Tabatinga I'd told him I could do a *haka,* a fierce Maori war dance like the one made famous by the New Zealand national rugby team, which performs it before each international match to intimidate its opponents. "If you do a haka for the Korubo, it'll help them accept you," he says to me now.

Led by Maya, the Korubo line up outside the maloca with puzzled expressions as I explain that I'm about to challenge one of their warriors to a fight — but, I stress, just in fun. After Possuelo tells them this is a far-off tribe's ritual before battle, Shishu, Maya's husband, steps forward to accept the challenge. I gulp nervously and then punch my chest and stamp my feet while screaming a bellicose chant in Maori. Jumi translates the words. "I die, I die, I live, I live." I stomp to within a few inches of Shishu, poke out my tongue Maori-style, and twist my features into a grotesque mask. He stares hard at me and stands his ground, refusing to be bullied. As I shout louder and punch my chest and thighs harder, my emotions are in a tangle. I want to impress the warriors with my ferocity but can't help fearing that if I stir them up, they'll attack me with their clubs.

I end my haka by jumping in the air and shouting, "Hee!" To my relief, the Korubo smile widely, apparently too practiced in real warfare to feel threatened by an unarmed outsider shouting and pounding his flabby chest. Possuelo puts an arm around my shoulder. "We'd better leave now," he says. "It's best not to stay too long on the first visit."

The next morning we return to the maloca, where Ta'van and other warriors have painted their bodies scarlet and flaunt headbands and armbands made from raffia streamers. Possuelo is astonished, never having seen them in such finery before. "They've done it to honor your haka," he says with a grin.

Shishu summons me inside the maloca. Jumi, rifle at the ready, follows. The low, narrow entrance — a precaution against a surprise attack — forces me to double over. As my eyes adjust to the dim light, I see the Korubo sprawled in vine hammocks strung low between poles holding up the roof or squatting by small fires. Stacked overhead on poles running the length of the hut are long slender blowpipes; axes and woven-leaf baskets lean against the walls. Holes dug into the dirt floor hold war clubs upright, at the ready. There are six small fireplaces, one for each family. Magna bustles about the hut, performing rudimentary medical checks and taking blood samples to test for malaria.

Maya, the hut's dominant presence, sits by a fireplace husking corn, which she'll soon begin grinding into mash. She hands me a grilled cob; delicious. Even the warriors are cooking and cleaning: muscular Teun sweeps the hut's earthen floor with a switch of tree leaves while Washman supervises. Tatchipan, a seventeen-year-old warrior who took part in the massacre of the white men, squats over a pot cooking the skinned carcass of a monkey. Ta'van helps his wife, Monan, boil a string of fish he'd caught in the river.

"The Korubo eat very well, with very little fat or sugar," says Magna. "Fish, wild pig, monkeys, birds, and plenty of fruit, manioc, and maize. They work hard and have a healthier diet than most Brazilians, so they have long lives and very good skin." Apart from battle wounds, the most serious illness they suffer is malaria, brought to the Amazon by outsiders long ago.

The men squat in a circle and wolf down the fish, monkey, and corn. Ta'van breaks off one of the monkey's arms complete with tiny hand and gives it to Tatchipan, who gnaws the skimpy meat from the bone. Even as they eat, I remain tense, worried they could erupt into violence at any moment. When I mention my concerns to Magna, whose monthly medical visits have given her a peek into the clan members' lives unprecedented for an outsider, she draws attention to their gentleness, saying, "I've never seen them quarrel or hit their children."

But they do practice one chilling custom: like other Amazon Indians, they sometimes kill their babies. "We've never seen it happen, but they've told us they do it," Magna says. "I know of one case where they killed the baby two weeks after birth. We don't know why."

Once past infancy, children face other dangers. Several years ago, Maya and her five-year-old daughter, Nwaribo, were bathing in the river when a massive anaconda seized the child, dragging her underwater. She was never seen again. The clan built a hut at the spot, and several of them cried day and night for seven days.

After the warriors finish eating, Shishu suddenly grips my arm, causing my heart to thump in terror. "You are *nowa,* a white man," he says. "Some nowa are good, but most are bad." I glance anxiously at Ta'van, who stares at me without expression while cradling his war club. I pray that he considers me one of the good guys.

Shishu grabs a handful of red *urucu* berries and crushes them between his palms, then spits into them and slathers the bloody-looking liquid on my face and arms. Hunching over a wooden slab studded with monkey teeth, he grinds a dry root into powder, mixes it with water, squeezes the juice into a coconut shell and invites me to drink. Could it be poison? I decide not to risk angering him by refusing it, and smile my thanks. The muddy liquid turns out to have an herbal taste, and I share several cups with Shishu. Once I'm sure it won't kill me, I half expect it to be a narcotic like kava, the South Seas concoction that also looks like grubby water. But it has no noticeable effect.

Other Korubo potions are not as benign. Later in the day Tatchipan places on a small fire by the hut's entrance a bowl brimming with curare, a black syrup that he makes by pulping and boiling a woody vine. After stirring the bubbling liquid, he dips the tips of dozens of slender blowpipe darts into it. The curare, Shishu tells me, is used to hunt small prey like monkeys and birds; it's not used on humans. He points to his war club, nestled against his thigh, and then his head. I get the message.

As the sun goes down, we return to Possuelo's base; even Possuelo, who the clan trusts more than any other white man, considers it too dangerous to stay overnight in the maloca. Early the next morning we're back, and they ask for the Maori war dance again. I comply, this time flashing my bare bottom at the end as custom demands. It may be the first time they've ever seen a white man's bum, and they roar with laughter at the sight. Still chuckling, the women head for the nearby maize and manioc fields. Shishu, meanwhile, hoists a twelve-foot-long blowpipe on his shoulder and strings a bamboo quiver, containing dozens of curare darts, around his neck. We leave the clearing together, and I struggle to keep up with him as he lopes through the shadowy jungle, alert for prey.

Hour slips into hour. Suddenly he stops and shades his eyes while peering up into the canopy. I don't see anything except tangled leaves and branches, but Shishu has spotted a monkey. He takes a dab of a gooey red ocher from a holder attached to his quiver and shapes it around the back of the dart as a counterweight. Then he takes the petals of a white flower and packs them around the ocher to smooth the dart's path through the blowpipe.

He raises the pipe to his mouth and, aiming at the monkey, puffs his cheeks and blows, seemingly with little effort. The dart hits the monkey square in the chest. The curare, a muscle relaxant that causes death by asphyxiation, does its work, and within several minutes the monkey, unable to breathe, tumbles to the forest floor. Shishu swiftly fashions a jungle basket from leaves and vine and slings the monkey over a shoulder.

By the end of the morning, he'll kill another monkey and a large, black-feathered bird. His day's hunting done, Shishu heads back to the maloca, stopping briefly at a stream to wash away the mud from his body before entering the hut.

Magna is sitting on a log outside the maloca when we return. It's a favorite spot for socializing: "The men and women work hard for about four or five hours a day and then relax around the maloca, eating, chatting, and sometimes singing," she says. "It'd be an enviable life except for the constant tension they feel, alert for a surprise attack even though their enemies live far away."

I see what she means later that afternoon, as I relax inside the maloca with Shishu, Maya, Ta'van, and Monan, the clan's friendliest woman. Their voices tinkle like music as we men sip the herbal drink and the women weave baskets. Suddenly Shishu shouts a warning and leaps to his feet. He's heard a noise in the forest, so he and Ta'van grab their war clubs and race outside. Jumi and I follow. From the forest we hear the familiar password, "Eh-heh," and moments later Tatchipan and another clan member, Marebo, stride into the clearing. False alarm.

Next morning, after I've performed the haka yet again, Maya hushes the noisy warriors and sends them out to fish in dugouts. Along the river they pull in to a sandy riverbank and begin to move along it, prodding the sand with their bare feet. Ta'van laughs with delight when he uncovers a buried cache of tortoise eggs, which he scoops up to take to the hut. Back on the river, the warriors cast vine nets and quickly haul up about twenty struggling fish, some shaded green with stumpy tails, others silvery with razor-sharp teeth: piranha. The nutritious fish with the bloodthirsty reputation is a macabre but apt metaphor for the circle of life in this feisty paradise, where hunter and hunted often must eat and be eaten by each other to survive.

In this jungle haunted by nightmarish predators, animal and human, the Korubo surely must also need some form of religion or spiritual practice to feed their souls as well as their bellies. But at the maloca I've seen no religious carvings, no rain-forest altars the Korubo might use to pray for successful hunts or other godly gifts. Back at the base that night, as Jumi sweeps a powerful searchlight back and forth across the river looking for intruders from downriver, Magna tells me that in the two years she's tended to clan members, she's never seen any evidence of their spiritual practice or beliefs. But we still know too little about them to be sure.

The mysteries are likely to remain. Possuelo refuses to allow anthropologists to observe the clan members firsthand, because, he says, it's too dangerous to live among them. And one day, perhaps soon, the clan will melt back into the deep jungle to rejoin a larger Korubo group. Maya and her clan broke away a decade ago, fleeing toward the river after warriors fought over her. But the clan numbers just twenty-three people, and some of the children are approaching puberty. "They've told me they'll have to go back to the main group one day to get husbands and wives for the young ones," says Magna. "Once that happens, we won't see them again." Because the larger group, which Possuelo estimates to be about one hundred fifty people, lives deep enough in the jungle's exclusion zone that settlers pose no threat, he's never tried to make contact with it.

Possuelo won't bring pictures of the outside world to show the Korubo, because he's afraid the images will encourage them to try to visit white settlements down the river. But he does have photographs he's taken from a small airplane of huts of still uncontacted tribes farther back in the Javari Valley, with as few as thirty people in a tribe and as many as four hundred. "We don't know their tribal names or languages, but I feel content to leave them alone because they're happy, hunting, fishing, farming, living their own way, with their unique vision of the world. They don't want to know us."

Is Sydney Possuelo right? Is he doing the isolated tribes of Brazil any favors by keeping them bottled up as premodern curiosities? Is ignorance really bliss? Or should Brazil's government throw open the doors of the twenty-first century to them, bringing them medical care, modern technology, and education? Before I left Tabatinga to visit the Korubo, the local Pentecostal church's Pastor Antonio, whose stirring sermons attract hundreds of the local Ticuna Indians, took Possuelo to task. "Jesus said, 'Go to the world and bring the Gospel to all peoples,'" Pastor Antonio told me. "The government has no right to stop us from entering the Javari Valley and saving the Indians' souls."

His view is echoed by many church leaders across Brazil. The resources of the exclusion zones are coveted by people with more worldly concerns as well, and not just by entrepreneurs salivating over the timber and mineral resources, which are worth billions of dollars. Two years ago more than five thousand armed men from the country's landless-workers movement marched into a tribal exclusion zone southeast of the Javari Valley, demanding to be given the land and sparking FUNAI officials to fear that they would massacre the Indians. FUNAI forced their retreat by threatening to call in the military.

But Possuelo remains unmoved. "People say I'm crazy, unpatriotic, a Don Quixote," he tells me when my week with the Korubo draws to a close. "Well, Quixote is my favorite hero because he was constantly trying to transform the bad things he saw into good." And so far, Brazil's political leaders have backed Possuelo.

As we get ready to leave, Ta'van punches his chest, imitating the haka, asking me to perform the dance one last time. Possuelo gives the clan a glimpse of the outside world by trying to describe an automobile. "They're like small huts that have legs and run very fast." Maya cocks her head in disbelief.

When I finish the war dance, Ta'van grabs my arm and smiles a farewell. Shishu remains in the hut and begins to wail, anguished that Possuelo is leaving. Tatchipan and Marebo, lugging war clubs, escort us down to the river.

The canoe begins its journey back across the millennia, and Possuelo looks back at the warriors, a wistful expression on his face. "I just want the Korubo and other isolated Indians to go on being happy," he says. "They have not yet been born into our world, and I hope they never are."

DANIEL ROTH

Torrential Reign

FROM *Fortune*

FOR TWO YEARS after the dot-com crash, Bram Cohen could almost always be found at his small dining room table, first in San Francisco's Nob Hill and later in Oakland. His long brown hair would flop in front of his eyes, and he'd curl it back over his ears as he stared at the screen of his Dell laptop, writing line after line after line of code. Occasionally Cohen would take breaks — there was a club to visit some nights, a conference on coding to help organize, a trip to Amsterdam — but then he'd return to his wooden chair, his keyboard on his lap, his laptop propped up on some books, his back perfectly straight (thanks to posture classes he was taking), and he'd program some more. First he lived off savings from the handful of jobs he'd had during the bubble. When that ran out, he lived off credit cards, following a rigid system for applying for and transferring debt to 0 percent introductory-rate cards. Friends would ask what he was doing. Why wouldn't he just get a job? Cohen shooed them away. He was determined to solve a puzzle that was consuming him.

Since the birth of the Internet, programmers had been stumped by how to transfer massive files — movies, TV shows, games, software, whatever — without incurring astronomical bills or risking frequent failure. Cohen knew he could find a solution; all it would take was time, good code, and brute intellect. He had all three. The money would take care of itself. "I didn't have any clear plans when I first started," he says. "I wasn't worried, partially because what I was doing was really cool, and partially because I'm broken and can't feel anxiety."

Cohen is not being self-deprecating. He never is. The thirty-year-old speaks in a disarmingly literal way about almost everything, including — and because of — his Asperger's syndrome. Often tagged as the "little-professor syndrome," the mild form of autism tends to give its sufferers superhuman abilities to concentrate on certain things but leaves them confused by very human social cues. "Even those individuals who have coped well with their handicap will strike one as strange," wrote one researcher. Cohen's condition is just bad enough that he has had to train himself to look people in the eye when they talk to him. But it has worked to his advantage, enabling him to obsessively turn over the downloading problem in his head.

What he came up with was BitTorrent, a deceptively simple program that has grown into the hottest way to download anything bigger than a music file — from the legal (like militaryvideos.net's amateur videos of the war in Iraq) to the infringing. It makes pirating a copy of the latest movie out of Hollywood a snap. All it takes is a free download of the BitTorrent software — something 45 million people have done — and anywhere from a few minutes to a few days. TorrentSpy, a site unrelated to Cohen that helps people find content available for download, averages more than six hundred new BitTorrent files a day. A sampling: Microsoft Office 2003, Alfred Hitchcock's *Rear Window,* episode two of CBS's *Ghost Whisperer* (in high definition, for serious Jennifer Love Hewittians), plus a file containing over four hundred *Amazing Spider-Man* scanned-in comics. Those huge files have made BitTorrent one of the biggest forces on the Internet, accounting for more than 20 percent of its traffic at any one time. That's double the volume generated by the most common Internet activities combined: clicking on Web pages, sending and receiving mail and spam, even streaming video clips.

With great power, of course, comes great enemies, so you can probably guess how it ought to play out. When the music-sharing networks Napster and Kazaa rose up earlier in this decade, the record labels sued them into submission. Surely BitTorrent will be next — especially now that Hollywood is beginning to feel the pinch as well. Today there are roughly 1.7 million copies of Hollywood movies — typically the most popular ones — being downloaded at any one time using BitTorrent, a 12 percent jump from last year, according to the online media measurement firm Big

Champagne. Analyst Informa Telecoms & Media estimates that in 2004 the downloads cost Hollywood roughly $860 million, or 4 percent of box office receipts. In the same period the number of TV shows downloaded grew by 150 percent — about 70 percent of them snagged using BitTorrent. "In the David and Goliath scenario, there really is a David," says BigChampagne's CEO, Eric Garland. "There's a kid at a keyboard who writes this incredibly disruptive technology."

Yet this time the plot has a twist: the entertainment industry seems to have found a disrupter it might be able to live with. In mid-September the recording industry issued cease-and-desist letters to seven popular downloading-technology companies, including BearShare, LimeWire, and eDonkey (prompting eDonkey's CEO to announce to a Senate subcommittee that he was "throwing in the towel"). BitTorrent was noticeably absent from the assault. It was also MIA last winter when the movie industry went after sites hosting copyright-busting BitTorrent content. Instead of fighting the entertainment establishment, Cohen is courting it. Last July he met Dan Glickman, president of the Motion Picture Association of America, for drinks at the Peninsula Hotel in Beverly Hills and left him wowed. "He's obviously a very brilliant guy," says Glickman, who notes that Hollywood understands it's time to embrace these new technologies. "The opportunities are going to be there to get our content to millions more people."

To understand how Cohen is managing to avoid Hollywood's wrath, you need to get inside his head. More inventor than entrepreneur, he has never claimed to be sticking it to the Man; nor, he insists, has he ever downloaded an infringing file. Napster's Shawn Fanning dreamed of changing the music industry. Cohen couldn't care less. True, in 1999 (pre-BitTorrent) Cohen noted on his Web site that he "build[s] systems to disseminate information, commit digital piracy, synthesize drugs, maintain untrusted contacts." He insists it was a parody of dot-commers' revolutionary thinking; the fact that BitTorrent operates without encryption or any attempt to hide its users' activity — downloaders beware! — back him up. Cohen says he created the program simply to . . . well, he's not sure why. "I wanted to work on something rewarding," he says. And once he was done, he was ready to move on to something new; his father had to twist his arm to build a company out of his work.

Last month the venture firm DCM-Doll Capital Management bet

that Cohen could indeed make BitTorrent a business, investing $8.75 million in the start-up. Now Cohen has to prove himself again, showing that he can thrive not just in the programming world — a place where logic rules and theories can be proved true or false — but in the fuzzy corporate world too, where compromise reigns and intellect doesn't always trump idiocy. "He just has to view the problem of how to be a CEO as yet another thing you can analyze and come to conclusions about," says his father, Barry Cohen, a computing sciences professor at the New Jersey Institute of Technology. "I don't think it's instinctive at all."

On a recent September day, Cohen is sitting in the lobby of the Marriott Wardman Park in Washington, D.C. His Merrell shoes are off, and his Gold-Toed feet are digging into the couch as he completes puzzles in *The Book of Sudoku #2* with a big orange ABC Family Channel pen. Hours earlier he keynoted a conference for the companies that control the backbone of the Internet. On stage Cohen sat with his legs crossed Indian-style — his shoes again placed neatly in front of him — as he explained the impact of BitTorrent. Now, instead of schmoozing with the attendees, he's happily digging into Sudoku: "It's very well balanced at a human level," he explains. "There is a decent-length list of tricks, where if you can do those reliably, you'll solve any Sudoku puzzle you come up against."

Cohen often delivers well-thought-out soliloquies on things that come to mind. Over breakfast one morning he weighed in on the difference between home fries and hash browns, why the telecom companies are nearly insolvent, and how the gas pedal of a car has evolved. (He's now trying to get his driver's license so that he can help with family chores; his wife is pregnant with their third child.) Each is an elegant summary of something that he has clearly picked over in his mind. The insights make his LiveJournal blog a must-read and are often followed by his staccato laugh, which arises from his thin frame without budging it at all. "Each topic you engage with him, he likes to go deep," says David Chao, cofounder of the DCM venture firm. "That's a classic sign of a great mind and a great engineer."

As Cohen was growing up on Manhattan's Upper West Side, his parents saw a great mind but knew that something was off. At six months he entertained himself for hours by simply staring at a wooden block and turning it over in his hands. The first words he learned to read — "goto," "run," and "print" — came from his family's Timex Sinclair computer, which he learned to program when he was five. It was clear that he excelled in certain areas, but the gap between Cohen and his peers kept growing. "He was always different — socially it wasn't quite as easy for him to make friends," says his younger brother, Ross, who is BitTorrent's chief technical officer and who is still not convinced that Bram has Asperger's. In fact, that diagnosis is relatively recent. The syndrome wasn't even detailed in the clinician's bible, *The Diagnostic and Statistical Manual of Mental Disorders,* until 1994, a year after Cohen graduated from the elite public Stuyvesant High School.

"I knew I was weird," says Cohen now. "I was pretty frustrated trying to interact with other people. I can really remember lots of stories in my life — things that it's really obvious to me now what was going on, but I didn't realize it back then because I didn't understand people very well." His confusion in social settings made people think he was a rebel or a slacker or both. He almost failed freshman math after he completed the first answer on a test and then refused to do the next forty-nine, declaring that they were simply variations on the first. When I called his school to find out about Cohen's time as cocaptain of the math team, the administrator who answered the phone blurted out about him, "What a space cadet!"

Not everyone reached such conclusions. In 1992 Cohen showed up uninvited at a New York University lecture given by Bart Selman, then a researcher at Bell Labs, on an arcane area of computer sciences. Cohen hung around afterward and grilled Selman with dead-on questions; Selman offered Cohen a summer internship typically reserved for college students and had him look at a new class of algorithm employed in software for heavy-duty problems like protein folding or complicated logistics. "We put him on it and he just went away, and a few weeks later he said, 'Okay, I've got a way to do this one hundred times faster,'" says Selman, now a professor at Cornell. "Many scientists — world-renowned people — had thought about it. Bram came in and did something new."

Unfortunately, MIT and other math and science schools cared less about Cohen's wizardry and more about his dismal grades. He enrolled at the University of Buffalo and dropped out two years later. Selman says he heard from Cohen only once, when the prodigy called to see whether Selman would write a recommendation for Cohen to work at RadioShack. Then the dot-com boom started, and Cohen, recently fired from a job at Kinko's for insubordination, found his way into one Web company after another. During that time he solved one of his biggest puzzles — why he acted the way he did — when he realized after much reading that he had Asperger's. By understanding why he did certain things and why others did what they did, Cohen was able to figure out how to get along better in the world. His wife, a former systems administrator whom he married in 2004, has even found an upside to his syndrome: "She actually finds my candor quite charming," he says. The next puzzle, the one that compelled Cohen to sequester himself, is the one that would finally make the world think of him not as a space cadet but as a star.

Before BitTorrent, large file transfers basically operated like the world's slowest Blockbuster. You found someone with, say, a movie or show you wanted by going on Kazaa or searching the Internet. Then you waited . . . and waited . . . and waited . . . as bit by orderly bit assembled itself on your PC, if it ever did. Cohen's brainstorm was to break the file into pieces — typically about one thousand — and share the pain of the transfer among many downloaders. The BitTorrent software runs on a user's machine and "talks" to other users who are trying to download the same file, automatically bartering for the pieces they each need. The more users, the faster the download. Cohen also filled the program with canny details. For example, when a file first goes up, machines can download chunks only as fast as they upload them, deterring freeloaders who want to receive but not give. The program also always tries to snag the rarest piece of the file first. The idea? The more machines that have that rare piece, the less rare it becomes.

The first real-world test of whether the principles would work on any large scale came in 2003, when the open-source software company Red Hat released its Red Hat Linux 9 operating system. Demand for the product was so strong that downloaders crippled Red Hat's servers. Eike Frost, a computer science student at Germany's University of Oldenburg, however, had managed to get a copy. He ran it through BitTorrent, then posted a link to the popular tech site Slashdot, inviting folks to come and get it. The swarm was immediate. Within three days the Red Hatters traded 21.15 terabytes of data — equivalent to more than all the books in the Library of Congress. At the peak, nearly 4,500 computers were swapping pieces of the file at any one time, uploading and downloading 1.4 terabytes each second. Frost estimates that if he had leased a line to handle that much traffic, it would have cost him $20,000 to $60,000. Instead he paid his usual $99 server hosting bill.

Pirates quickly saw the benefit, but so too did some legitimate enterprises. The gamemaker Blizzard Entertainment uses BitTorrent to distribute the two-gigabyte World of Warcraft game (nearly three CDs' worth of info) and all the patches that go with it. "We have hundreds of thousands, if not millions, getting our game," says Blizzard's COO, Paul Sams. Sun Microsystems is using BitTorrent to make available its entire Open Solaris operating system to tens of thousands of users and is planning to boost its BitTorrent use as it open-sources all its software over the next few years. Others, like the anime film giant ADV Films, have tapped BitTorrent as the best way to spread trailers of films to multitudes of fans at once.

Those are all fantastic votes of confidence, except for one thing: mone of the companies has to pay BitTorrent a penny. Cohen created his program under an open-source license, leaving anyone free to tinker with it, distribute it, or use it. "The only payments that BitTorrent accepts are completely optional donations," the BitTorrent site cheerfully notes. The capital infusion from DCM will pay the salaries of the company's thirteen employees for now, but it won't ensure the company's long-term survival. Cohen and the COO, Ashwin Navin, however, have a plan they think will build on the founder's brainstorm.

Navin runs through a verbal sketch of the media industry as he sees it. He's sitting in BitTorrent's office, which for now is a room in a converted warehouse near the old port of San Francisco. BitTorrent, explains the twenty-eight-year-old, has commoditized the most expensive part of the media equation: distribution. Now it has the chance to cash in on the content itself. The company plans to establish a marketplace — part iTunes, part eBay — for bandwidth-intensive content. BitTorrent will host and index any content its creators want to sell (or give away, for that matter). It will generate revenue either by charging sellers a small commission or through related advertising. Until it has built the marketplace,

BitTorrent is employing the ubiquitous business model of the Web 2.0 age: in partnership with AskJeeves, it's selling ads against a BitTorrent content search engine that it maintains on its home page.

Search ads? Online marketplace? The only thing missing is a podcast. BitTorrent was born of programming genius, but its business model seems secondhand. Plus it's late. Google this spring launched Google Video, which allows users to post streams (not downloads, as the BitTorrent site would permit) and, if they wish, sell their content. Even scarier, Google doesn't set size limits on the files it will host, making BitTorrent's cheap and simple handling of large files more academic than vital. In mid-October Apple unveiled its long-rumored video iPod and started making some TV downloads and Pixar shorts available through its popular iTunes service. Navin says that the Google and Apple moves are both competition but that BitTorrent's market will offer much more than just movies and TV shows. Also, he speculates that Apple is paying "an astronomical price for bandwidth."

Faced with such fearsome competition, the challenge for BitTorrent isn't one of engineering but of business: it needs to set its service apart. So this summer it retained Fred Davis, a high-profile entertainment lawyer who is the son of recording industry legend Clive Davis, to handle discussions with the music industry about securing licenses that would allow BitTorrent to sell everything from albums to videos to high-fidelity tracks that would be too big to simply stick on a CD. "Everybody knows BitTorrent," says Davis. "When I call up and say, 'Let's meet — they want to be legitimate,' we are welcomed at the front door of every company." Navin hints that similar discussions are in the works with Hollywood.

Cohen realizes that it's going to take someone like Navin to keep the company running — Navin's personable to a fault, while Cohen has to work at it — but he doesn't want to have too loose a grip. "Right now I'm the CEO because I don't trust anyone else to be the CEO," he says. To keep things interesting, he focuses on trying to predict the future and on approaching the whole affair as one more puzzle to master. "I really hope to make it possible for a whole lot of content creators to take their stuff online and to make money off it," says Cohen, listing both Google and eBay as corporate role models. Creating such a market would be good for both the creators and for his company, he notes. "But I'm really more proud of making money for other people than for myself. Making money for oneself is not widely considered a high moral goal, though it's an acceptable one."

Almost no one in media or entertainment doubts that the world will move toward digital distribution. And as bandwidth costs drop and the media powerhouses get more tech-savvy, the desire to control the pipe to their consumers — middleman be damned — is just too tempting. In one of his last speeches as a Hollywood don, Michael Eisner in late September told his peers: "Don't panic over the latest techno jargon, such as peer-to-peer, Wi-Max, 802.11, BitTorrent. Rather, embrace them." So will BitTorrent walk into the promised land with the moviemakers? Well, that's another matter.

So far Hollywood has given BitTorrent only a deferred sentence. That's enough for some. Certainly the company's investors are crossing their fingers for something Skype-like to happen: the infant Internet-based phone company was recently purchased by eBay for up to $4.1 billion. "I'm a big believer that when the majority of Internet traffic is governed by BitTorrent and they have forty-five million users, you're going to be able to monetize that," says DCM's Chao.

But it's far from sure that Cohen's going to stick around to guide the process. Henry Kautz, who worked closely with Cohen at Bell Labs and is now a professor at the University of Washington, isn't convinced that Cohen would mind all that much if BitTorrent doesn't pay off. Last fall the former intern dropped by Kautz's office unannounced to quiz him on an area of computer science called satisfiability testing. The two also got a chance to catch up on life since Bell Labs. Cohen talked about BitTorrent, but he didn't dwell on it, leaving Kautz with a strange feeling. "If BitTorrent was outlawed and went away, I don't think it would hurt him emotionally at all," he says. "He looks at it as a fascinating puzzle that he's figured out."

I told Cohen what Kautz had said, and he paused to think about it, resting his Sudoku book on his lap. "It wouldn't leave me emotionally scarred," he says. "I mean, I'm certainly going to try to keep that from happening. But if for some reason the shit hits the fan, you just deal with it. That's the way I've always been." There are, after all, plenty more puzzles out there to be solved.

JESSICA SNYDER SACHS

Are Antibiotics Killing Us?

FROM *Discover*

ALAN HUDSON LIKES TO TELL A STORY about a soldier and his high school sweetheart. The young man returns from an overseas assignment for their wedding with a clean bill of health, having dutifully cleared up an infection of sexually transmitted chlamydia.

"Three weeks later, the wife has a screaming genital infection," Hudson recounts, "and I get a call from the small-town doctor who's trying to save their marriage." The soldier, it seems, has decided his wife must have been seeing other men, which she denies.

Hudson pauses for effect, stretching back in his seat and propping his feet on an open file drawer in a crowded corner of his microbiology laboratory at Wayne State Medical School in Detroit. "The doctor is convinced she's telling the truth," he continues, folding his hands behind a sweep of white, collar-length hair. "So I tell him, 'Send me a specimen from him and a cervical swab from her.'" This is done after the couple has completed a full course of antibiotic treatment and tested free of infection.

"I PCR 'em both," Hudson says, "and he is red hot."

PCR stands for polymerase chain reaction — a technique developed about twenty years ago that allows many copies of a DNA sequence to be made. It is often used at crime scenes, where very little DNA may be available. Hudson's use of the technique allowed him to find traces of chlamydia DNA in the soldier and his wife that traditional tests miss because the amounts left after antibiotic treatment are small and asymptomatic. Nonetheless, if a small number of inactive chlamydia cells passed from groom to bride, the infection could have became active in its new host.

Hudson tells the tale to illustrate how microbes that scientists once thought were easily eliminated by antibiotics can still thrive in the body. His findings and those of other researchers raise disturbing questions about the behavior of microbes in the human body and how they should be treated.

For example, Hudson has found that quiescent varieties of chlamydia may play a role in chronic ailments not traditionally thought to be related to this infectious agent. In the early 1990s he found two types of chlamydia — *Chlamydia trachomatis* and *Chlamydia pneumonia* — in the joint tissue of patients with inflammatory arthritis. More famously, in 1996, he began fishing *C. pneumonia* out of the brain cells of Alzheimer's victims. Since then other researchers have made headlines after reporting the genetic fingerprints of *C. pneumonia,* as well as several kinds of common mouth bacteria, in the arterial plaque of heart attack patients. Hidden infections are now thought to be the basis of still other stubbornly elusive ills like chronic fatigue syndrome, Gulf War syndrome, multiple sclerosis, lupus, Parkinson's disease, and types of cancer.

To counteract these killers, some physicians have turned to lengthy or lifelong courses of antibiotics. At the same time, other researchers are counterintuitively finding that bacteria we think are bad for us also ward off other diseases and keep us healthy. Using antibiotics to tamper with this complicated and little-understood population could irrevocably alter the microbial ecology in an individual and accelerate the spread of drug-resistant genes to the public at large.

The two-faced puzzle regarding the role of bacteria is as old as the study of microbiology itself. Even as Louis Pasteur became the first to show that bacteria can cause disease, he assumed that bacteria normally found in the body are essential to life. Yet his protégé, Élie Metchnikoff, openly scoffed at the idea. Metchnikoff blamed indigenous bacteria for senility, atherosclerosis, and an altogether shortened life span — going even so far as to predict the day when surgeons would routinely remove the human colon simply to rid us of the "chronic poisoning" from its abundant flora.

Today we know that trillions of bacteria carpet not only our intestines but also our skin and much of our respiratory and urinary tracts. The vast majority of them seem to be innocuous, if not beneficial. And bacteria are everywhere, in abundance — they outnumber other cells in the human body by ten to one. David Relman and his team at Stanford University and the Veterans Administration Medical Center in Palo Alto, California, recently found the genetic fingerprints of several hundred new bacterial species in the mouths, stomachs, and intestines of healthy volunteers.

"What I hope," Relman says, "is that by starting with specimens from healthy people, the assumption would be that these microbes have probably been with us for some time relative to our stay on this planet and may, in fact, be important to our health."

Meanwhile, the behavior of even well-known bacterial inhabitants is challenging the old, straightforward view of infectious disease. In the nineteenth century Robert Koch laid the foundation for medical microbiology, postulating that any microorganism that causes a disease should be found in every case of the disease and always cause the disease when introduced into a new host. That view prevailed until the middle of this past century. Now we are more confused than ever. Take *Helicobacter pylori.* In the 1980s infection by the bacterium, not stress, was found to be the cause of most ulcers. Overnight, antibiotics became the standard treatment. Yet in the undeveloped world ulcers are rare, and *H. pylori* is pervasive.

"This stuff drives the old-time microbiologists mad," says Hudson, "because Koch's postulates simply don't apply." With new technologies like PCR, researchers are turning up stealth infections everywhere, yet they cause problems only in *some* people *sometimes,* often many years after the infection.

These mysteries have nonetheless not stopped a free flow of prescriptions. Many rheumatologists, for example, now prescribe long-term — even lifelong — courses of antibiotics for inflammatory arthritis, even though it isn't known if the antibiotics actually clear away bacteria or reduce inflammatory arthritis in some other unknown manner.

Even more far-reaching is the use of antibiotics to treat heart disease, a trend that began in the early 1990s after studies associated *C. pneumonia* with the accumulation of plaque in arteries. In April two large-scale studies reported that use of antibiotics does not reduce the incidence of heart attacks or eliminate *C. pneumonia.* But researchers left antibiotic-dosing cardiologists a strange option by admitting that they do not know if stronger, longer courses of antibiotics or combined therapies would succeed.

Meanwhile, many researchers are alarmed. Curtis Donskey, an infectious-diseases specialist at Case Western Reserve University in Cleveland, says, "Unfortunately, far too many physicians are still thinking of antibiotics as benign. We're just now beginning to understand how our normal microflora does such a good job of preventing our colonization by disease-causing microbes. And from an ecological point of view, we're just starting to understand the medical consequences of disturbing that with antibiotics."

Donskey has seen the problem firsthand at the Cleveland's VA Medical Center, where he heads infection control. "Hospital patients get the broadest spectrum, most powerful antibiotics," he says, but they are also "in an environment where they get exposed to some of the nastiest, most drug-resistant pathogens." Powerful antibiotics can be dangerous in such a setting because they kill off harmless bacteria that create competition for drug-resistant colonizers, which can then proliferate. The result: hospital-acquired infections have become a leading cause of death in critical-care units.

"We also see serious problems in the outside community," Donskey says, because of inappropriate antibiotic use. The consequences of disrupting the body's bacterial ecosystem can be minor, such as a yeast infection, or they can be major, such as the overgrowth of a relatively common gut bacterium called *Clostridium difficile.* A particularly nasty strain of *C. difficile* has killed hundreds of hospital patients in Canada over the past two years. Some had checked in for simple, routine procedures. The same strain is moving into hospitals in the United States and the United Kingdom.

Jeffrey Gordon, a gastroenterologist turned full-time microbiologist, heads the spanking new Center for Genomic Studies at Washington University in Saint Louis. The expansive, sun-streaked laboratory sits above the university's renowned gene-sequencing center, which was a major player in powering the Human Genome Project. "Now it's time to take a broader view of the human genome," says Gordon, "one that recognizes that the human body probably contains one hundred times more microbial genes than human ones."

Gordon supervises a lab of some twenty graduate students and postdocs with expertise in disciplines ranging from ecology to crystallography. Their collaborations revolve around studies of unusually successful colonies of genetically engineered germ-free mice and zebra fish.

Gordon's veteran mouse wranglers, Marie Karlsson and her husband, David O'Donnell, manage the rearing of germ-free animals for comparison with genetically identical animals that are colonized with one or two select strains of normal flora. In a cavernous facility packed with rows of crib-size bubble chambers, Karlsson and O'Donnell handle their germ-free charges via bulbous black gloves that serve as airtight portals into the pressurized isolettes. They generously supplement sterilized mouse chow with vitamins and extra calories to replace or complement what is normally supplied by intestinal bacteria. "Except for their being on the skinny side, we've got them to the point where they live near-normal lives," says O'Donnell. Yet the animals' intestines remain thin and underdeveloped in places, bizarrely bloated in others. They also prove vulnerable to any stray pathogen that slips into their food, water, or air.

All of Gordon's protégés share an interest in following the molecular crosstalk among resident microbes and their host when they add back a component of an animal's normal microbiota. One of the most interesting players is *Bacteroides thetaiotaomicron,* or *B. theta,* the predominant bacterium of the human colon and a particularly bossy symbiont.

The bacterium is known for its role in breaking down otherwise indigestible plant matter, providing up to 15 percent of its host's calories. But Gordon's team has identified a suite of other, more surprising skills. Three years ago they sequenced *B. theta*'s entire genome, which enabled them to work with a gene chip that detects what proteins are being made at any given time. By tracking changes in the activity of these genes, the team has shown that *B. theta* helps guide the normal development and functioning of the intestines — including the growth of blood vessels, the proper turnover of epithelial cells, and the marshaling of components of the immune system needed to keep less well behaved bacteria at bay. *B. theta* also exerts hormonelike long-range effects that may help the host weather times when food is scarce and ensure the bacterium's own survival.

Fredrik Bäckhed, a young postdoc who came to Gordon's laboratory from the Karolinska Institute in Stockholm, has caught *B. theta* sending biochemical messages to host cells in the abdomen, directing them to store fat. When he gave germ-free mice an infusion of gut bacteria from a conventionally raised mouse, they immediately put on an average of 50 percent more fat, even though they were consuming 30 percent less food than when they were germ-free. "It's as if *B. theta* is telling its host, 'Save this — we may need it later,'" Gordon says.

Justin Sonnenburg, another postdoctoral fellow, has documented that *B. theta* turns to the host's body for food when the animal stops eating. He has found that when a lab mouse misses its daily ration, *B. theta* consumes the globs of sugary mucus made every day by some cells in the intestinal lining. The bacteria graze on these platforms, which the laboratory has dubbed Whovilles (after the dust-speck metropolis of Dr. Seuss's *Horton Hears a Who!*). When the host resumes eating, *B. theta* returns to feeding on the incoming material.

Gordon's team is also looking at the ecological dynamics that take place when combinations of normal intestinal bacteria are introduced into germ-free animals. And he plans to study the dynamics in people by analyzing bacteria in fecal samples.

Among the questions driving him: Can we begin to use our microbiota as a marker of health and disease? Does the makeup of this "bacterial nation" shift when we become obese, try to lose weight, experience prolonged stress, or simply age? Do people in Asia or Siberia harbor the same organisms in the same proportions as those in North America or the Andes?

"We know that our environment affects our health to an enormous degree," Gordon says. "And our microbiota are our most intimate environment by far."

A couple hundred miles northeast of Gordon's laboratory, microbiologist Abigail Salyers, at the University of Illinois at Urbana-Champaign, has been exploring a more sinister feature of our bacteria and their role in antibiotic resistance. At the center of her research stands a room-size walk-in artificial "gut" with the thermostat set at the human intestinal temperature of 100.2 degrees Fahrenheit. Racks of bacteria-laced test tubes line three walls, the sealed vials purged of oxygen to simulate the anaerobic conditions inside a colon. Her study results are alarming.

Salyers says her research shows that decades of antibiotic use have bred a frightening degree of drug resistance into our intestinal flora. The resistance is harmless as long as the bacteria remain confined to their normal habitat. But it can prove deadly when those bacteria contaminate an open wound or cause an infection after surgery.

"Having a highly antibiotic-resistant bacterial population makes a person a ticking time bomb," says Salyers, who studies the genus *Bacteroides*, a group that includes not only *B. theta* but also about a quarter of the bacteria in the human gut. She has tracked dramatic increases in the prevalence of several genes and suites of genes coding for drug resistance. She's particularly interested in tetQ, a DNA sequence that conveys resistance to tetracycline drugs.

When her team tested fecal samples taken in the 1970s, they found that less than 25 percent of human-based *Bacteroides* carried tetQ. By the 1990s that rate had passed the 85 percent mark, even among strains isolated from healthy people who hadn't used antibiotics in years. The dramatic uptick quashed hopes of reducing widespread antibiotic resistance by simply withdrawing or reducing the use of a given drug. Salyers's team also documented the spread of several *Bacteroides* genes conveying resistance to other antibiotics, such as macrolides, which are widely used to treat skin, respiratory, genital, and blood infections.

As drug-resistant genes become common in bacteria in the gut, they are more likely to pass on their information to truly dangerous bugs that move only periodically through our bodies, says Salyers. Even distantly related bacteria can swap genes with one another using a variety of techniques, from direct cell-to-cell transfer, called conjugation, to transformation, in which a bacterium releases snippets of DNA that other bacteria pick up and use.

"Viewed in this way, the human colon is the bacterial equivalent of eBay," says Salyers. "Instead of creating a new gene the hard way — through mutation and natural selection — you can just stop by and obtain a resistance gene that has been created by some other bacterium."

Salyers has shown that *Bacteroides* probably picked up erythromycin-resistant genes from distantly related species of staphylococcus and streptococcus. Although neither bug colonizes the intestine, they are routinely inhaled and swallowed, providing a window of twenty-four to forty-eight hours in which they can commingle with intestinal flora before exiting. "That's more than long enough to pick up something interesting in the swinging singles bar of the human colon," she quips.

Most disturbing is Salyers's discovery that antibiotics like tetracycline actually stimulate *Bacteroides* to begin swapping its resistance genes. "If you think of the conjugative transfer of resistance genes as bacterial sex, you have to think of tetracycline as the aphrodisiac," she says. When Salyers exposes *Bacteroides* to other bacteria, such as *Escherichia coli,* under the disinhibiting influence of antibiotics, she has witnessed the step-by-step process by which the bacteria excise and transfer the tetQ gene from one species to another.

Nor is *Bacteroides* the only intestinal resident with such talents. "In June 2002 we passed a particularly frightening milestone," Salyers says. That summer epidemiologists discovered hospital-bred strains of the gut bacterium *Enterococcus* harboring a gene that made them impervious to vancomycin. The bacterium may have since passed the gene to the far more dangerous *Staphylococcus aureus,* the most common cause of fatal surgical and wound infections.

"I am completely mystified by the lack of public concern about this problem," she says. With no simple solution in sight, Salyers continues to advise government agencies such as the Food and Drug Administration and the Department of Agriculture to reduce the use of antibiotics in livestock feed, a practice banned throughout the European Union. She supports the prescient efforts of the Tufts University microbiologist Stuart Levy, founder of the Alliance for the Prudent Use of Antibiotics, which has been hectoring doctors to use antibiotics more judiciously.

Yet just when the message appears to be getting through — judging by a small but real reduction in antibiotic prescriptions — others are calling for an unprecedented increase in antibiotic use to clear the body of infections we never knew we had. Among them is William Mitchell, a Vanderbilt University chlamydia specialist. If antibiotics ever do prove effective for treating coronary artery disease, he says, the results would be "staggering. We're talking about the majority of the population being on long-term antibiotics, possibly multiple antibiotics."

Hudson cautions that before we set out to eradicate our bacterial fellow travelers, "we'd damn well better understand what they're doing in there." His interest centers on chlamydia, with its maddening ability to exist in inactive infections that flare into problems only for an unlucky few. Does the inactive form cause damage by secreting toxins or killing cells? Or is the real problem a disturbed immune response to them?

Lately Hudson has resorted to a device he once shunned in favor of DNA probes: a microscope, albeit an exotic $250,000 model. This instrument, which can magnify organisms an unprecedented 15,000 times, sits in the laboratory of Hudson's spouse, Judith Whittum-Hudson, a Wayne State immunologist who is working on a chlamydia vaccine. On a recent afternoon Hudson marveled as a shimmering chlamydia cell was beginning to morph from its infectious stage into its mysterious and bizarre-looking persistent form. "One minute you have this perfectly normal, spherical bacterium and the next you have this big, goofy-looking doofus of a microbe," he says. He leans closer, focusing on a roiling spot of activity. "It's doing something. It's making something. It's saying something to its host."

OLIVER SACKS

Remembering Francis Crick

FROM *The New York Review of Books*

1

I READ THE FAMOUS "double helix" letter by James Watson and Francis Crick in *Nature* when it was published in 1953 — I was an undergraduate at Oxford then, reading physiology and biochemistry. I would like to say that I immediately saw its tremendous significance, but this was not the case for me or, indeed, for most people at the time.

It was only in 1962, when Francis Crick came to talk at Mount Zion Hospital in San Francisco, where I was interning, that I started to realize the vast implications of the double helix. Crick's talk at Mount Zion was not on the configuration of DNA but on the work he had been doing with the molecular biologist Sidney Brenner to determine how the sequence of DNA bases could specify the amino acid sequence in proteins. They had just shown, after four years of intense work, that the translation involved a three-nucleotide code. This was itself a discovery no less momentous than the discovery of the double helix.

But Crick's mind was always moving forward, and clearly he had already moved on to other things. There were, he intimated in his talk, two "other things," great enterprises whose exploration lay in the future: understanding the origin and nature of life and understanding the relation of brain and mind — in particular, the biological basis of consciousness. Did he have any inkling, any conscious thought, when he spoke to us in 1962, that these would be the very subjects he himself would address in the years to come, once he had "dealt with" molecular biology, or at least taken it to the stage where it could be delegated to others?

It was not until May of 1986 that I met Francis Crick, at a conference in San Diego. There was a big crowd, full of neuroscientists, but when it was time to sit down for dinner, Crick singled me out, seized me by the shoulders, sat me down next to him, and said, "Tell me stories!" I have no memory of what we ate, or anything else about the dinner, only that I told him stories about many of my patients, and that each one set off bursts of hypotheses, theories, suggestions for investigation in his mind. Writing to Crick a few days later, I said that the experience was "a little like sitting next to an intellectual nuclear reactor . . . I never had a feeling of such *incandescence.*"

He was especially eager to hear stories of visual perception, and was fascinated when I told him of a patient who had consulted me a few weeks before, an artist who had experienced a sudden and total loss of color perception following a car accident (his loss of color vision was accompanied by an inability to visualize or to dream in color). Crick was also fascinated when I told him how a number of my migraine patients had experienced, in the few minutes of a migraine aura, a flickering of static, "frozen" images in place of their normal, continuous visual perception. He asked me whether such "cinematic vision" (as I called it) was ever a permanent condition, or one that could be elicited in a predictable way so that it could be investigated. I said I did not know.

During 1986, encouraged by the questions Crick had fired at me, I spent a good deal of time with my colorblind patient, Mr. I., and in January of 1987, I wrote to Crick. "I have now written up a longish report on my patient . . . Only in the actual writing did I come to see how color might indeed be a (cerebro-mental) construct." I had now started to wonder, I added, whether all perceptual qualities, including the perception of motion, were similarly constructed by the brain. I had spent most of my professional life wedded to notions of "naive realism," regarding visual perceptions, for example, as mere transcriptions of retinal images; that was very much the epistemological atmosphere at the time. But now, as I worked with Mr. I., this was giving way to a very different vision of the brain-mind, a vision of it as essentially constructive or creative.

(I also included a copy of my book *A Leg to Stand On,* because it contained accounts of personal experiences of both motion and depth blindness.)

I got a letter back a few days later — Crick was the promptest of correspondents — in which he sought more detail about the difference between my migraine patients and a remarkable motion-blind patient described by the German neuropsychologist Josef Zihl. My migraine patients experienced "stills" in rapid succession, whereas in Zihl's patient (who had acquired motion blindness following a stroke), the stills apparently lasted much longer, perhaps several seconds each. In particular, Crick wanted to know whether in my patients successive stills occurred within the interval between successive eye movements or only between such intervals. "I would very much like to discuss these topics with you," he wrote, "including your remarks about color as a cerebro-mental construct."

In my reply to Crick's letter, I enlarged on the deep differences between my migraine patients and Zihl's motion-blind woman. I mentioned too that I was working on Mr. I.'s case with several colleagues. My ophthalmologist friend Bob Wasserman had examined Mr. I. several times and discussed his case with me at length, and we had also been joined by a young neuroscientist, Ralph M. Siegel, who had just come to New York from the Salk Institute, where, as it happened, he had been close to Crick. Siegel collaborated with us, designing and conducting a variety of psychophysical experiments with our patient. I mentioned, too, that the neurophysiologist Semir Zeki had visited us from London and had tested our patient with his color "Mondrians," using light of different wavelengths, and with these had confirmed that Mr. I. showed excellent wavelength discrimination. His retinal cones were still reacting to light of different wavelengths, and his primary visual cortex (an area that has been named V1) was registering this information, but it was clear that this wavelength discrimination was not in itself sufficient for the experience of color. The actual perception of color had to be *constructed,* and constructed by an additional process in another part of the visual cortex. Zeki felt, on the basis of animal experiments, that color was constructed in the tiny areas of the visual cortex called V4, and wondered whether Mr. I. had damage to these areas resulting from his stroke.

At the end of October 1987, I was able to send Crick the paper that Bob Wasserman and I had written, "The Case of the Colorblind Painter," and having sent this article off to him, I was assailed by intense anxieties — what would he say? — but also by great eagerness to hear his reactions. Both of these feelings waxed in the weeks that followed, for Crick was out of town, his secretary told us, until mid-December.

Early in January, then, I got a response from Crick — an absolutely stunning letter, five pages of single-spaced typing, minutely argued and bursting with ideas and suggestions. Some of them, he said, were "wild speculation," but it was the sort of wildness that had intuited the double-helix structure of DNA thirty-five years before. "Do please excuse the length of this letter," he added. "We might talk about it over the phone, after you've had time to digest it all." Bob and I, and Ralph too, were mesmerized by the letter. It seemed to get deeper and more suggestive every time we read it, and we got the sense that it would need a decade or more of work, by a dedicated team of psychophysicists, neuroscientists, brain imagers, and others, to follow up on the torrent of suggestions Crick had made.

I wrote back to Crick, saying that we would need weeks or months to digest all he had said, but would be getting to work in the meantime, doing tests of motion vision, stereo vision, and contrast vision in Mr. I., as well as more sophisticated color-vision testing, and that we hoped to get high-quality MRI and PET scans to look at the activity in his visual cortex.

In his five-page letter, Crick had written, "Thank you so much for sending me your fascinating article on the color-blind artist . . . Even though, as you stress in your letter, it is not strictly a scientific article, it has aroused much interest among my colleagues and my scientific and philosophical friends here. We have had a couple of group sessions on it and in addition I have had several further conversations with individuals." He added that he had sent a copy of the article and his letter to David Hubel, who, with Torsten Wiesel, had done pioneering work on the cortical mechanisms of visual perception.

Writing to me again in January of 1988, he said:

> So glad to hear from you and to learn that you plan more work on Mr. I. All the things you mention are important, especially the scans . . . There

> is no consensus yet among my friends about what the damage might be in such cases of cerebral achromatopsia. I have (very tentatively) suggested the V1 blobs plus some subsequent degeneration at higher levels, but this really depends on seeing little in the scans (if most of V4 is knocked out you should see something). David Hubel tells me that he favors damage to V4, though this opinion is preliminary. David van Essen tells me that he suspects some area further upstream.

He mentioned two of Antonio Damasio's cases: in one of these, the patient had lost color imagery, but still dreamed in color. (She later regained her color vision.) "I think the moral of all this," he concluded, "is that only careful and extensive psychophysics on [such] a patient *plus* accurate localization of the damage will help us. (So far, we cannot see how to study visual imagery and dreams in a monkey.)"

I was very excited to think that Crick was opening our paper, our "case," for discussion in this way. It gave me a deeper sense of science as a communal enterprise, of scientists as a fraternal, international community, sharing and thinking on each other's work — and of Crick himself as a sort of hub, or center, in touch with everyone in this neuroscientific world.

Eighteen months passed without further contact between us, for I was largely occupied now with other, nonvisual, conditions. But in August of 1989, I wrote to Crick again, saying that I was still working with our colorblind patient. I also enclosed a copy of my just-published book, *Seeing Voices,* on sign language and the culture and history of congenitally deaf people. I was especially fascinated by the way in which novel perceptual and linguistic powers could develop in congenitally deaf people and by the brain changes that both resulted from and allowed a very different perceptual experience in them, and in my book I had cited some of Crick's thoughts in this context — he had just published a paper called "The Recent Excitement About Neural Networks." (I also cited Gerald M. Edelman in the same context, and wondered what the relationship of these two extraordinary figures might be, since Edelman had recently relocated his Neurosciences Institute to La Jolla, practically next door to the Salk Institute, where Crick was working.)

Crick wrote back a few days later and said that he had read my original articles about the deaf and American Sign Language in the

New York Review. He had become intrigued by the subject and looked forward to reading the book ("including the many fascinating footnotes"). "Over the years," he added, "Ursula [Bellugi, his colleague at the Salk] has been patiently educating me about ASL." He urged me, too, to continue my investigation of Mr. I. and enclosed the manuscript of a new article of his: "At the moment I am trying to come to grips with visual awareness, but so far it remains as baffling as ever."

The "short article" he enclosed, "Towards a Neurobiological Theory of Consciousness," was one of the first synoptic articles to come out of his collaboration with Christof Koch at Caltech. I felt very privileged to see this manuscript, in particular their carefully laid out argument that an ideal way of entering this seemingly inaccessible subject would be through exploring the mechanisms and disorders of visual perception.

Crick and Koch's paper covered a vast range in a few pages, was aimed at neuroscientists, and was sometimes dense and highly technical. But I knew that Crick could also write in a very accessible and witty and personable way — this was especially evident in his two earlier books, *Life Itself* (1981) and *Of Molecules and Men* (1966). So I now entertained hopes that he might give a more popular and accessible form to his neurobiological theory of consciousness, enriched with clinical and everyday examples. He intimated, in one of his letters, that he would attempt such a book, and in September 1993 his publishers sent me a proof of *The Astonishing Hypothesis.*

I read this at once with great admiration and delight, and wrote to Crick right away:

> I think you bring together an incredible range of observations from different disciplines into a single, brilliant clear focus . . . I am particularly and personally grateful that you make such a full and generous reference to the Colorblind Painter whom Bob Wasserman and I studied. I still cherish that marvellous letter you sent me about him. When Semir [Zeki] developed his new technique for PET scanning V4 etc. in humans, we did our utmost to get Mr. I. to him, but sadly, Mr. I. became acutely ill at this time with bronchogenic carcinoma and brain metastases, and died within a few weeks (we were not able to get a postmortem). So it never became clear exactly what happened.

I enclosed a copy of my new article, "To See and Not See," saying that here, too, Bob Wasserman and Ralph Siegel had been invaluable research collaborators and that we especially hoped to get serial PET scans, as this patient, Virgil (who was born virtually blind and given vision through eye surgery as an adult), struggled to establish basic visual perceptions in the visual tumult suddenly loosed on him. (Though here, too, the patient, like Mr. I., became unable, owing to an unrelated illness, to have such scans.) "I know your own central interests lie in vision, and the ways in which this can illuminate the fundamentals of mind and brain," I wrote, "and in my own rambling, clinical way, I find it is my own favorite subject, too."

The following year, in June 1994, I met with Ralph Siegel and Francis Crick for dinner in New York. As with that first dinner in 1986, I cannot remember what we ate, only that the talk ranged in all directions. Ralph talked about his current work with visual perception in monkeys and his thoughts on the fundamental role of chaos at the neuronal level (we had worked together writing about chaos and self-organization in the phenomena of visual migraines, as well as chaos in parkinsonism). Francis spoke about his expanding work with Koch and their latest theories about the neural correlates of consciousness, and I spoke about my upcoming visit to Pingelap, an island in the South Pacific with a genetically isolated population of people born completely colorblind — I planned to travel there with Bob Wasserman and a Norwegian perceptual psychologist, Knut Nordby, who, like the Pingelapese, had been born without color receptors in his retinas.

In February 1995 I sent Francis my new book, *An Anthropologist on Mars,* which contained an expanded version of "The Case of the Colorblind Painter," much amplified, in part, through my discussions with him on the case. (He had immediately assented to my quoting from some of his letters in the revised version, adding that he liked it even more than the original, "if only because you have conveyed more of Mr. I.'s personality.") I also told him something of my experiences in Pingelap, and how Knut and I tried to imagine what changes might have occurred in his brain in response to his achromatopsia. Would the color-constructing centers in his brain have atrophied in the absence of any color receptors in his retinas? Would his V4 areas have been reallocated for other visual functions? Or were they, perhaps, still awaiting an input, an input that might be provided by direct electrical or magnetic stimulation? And if this could be done, would he, for the first time in his life, see color? Would he know it *was* color, or would this visual experience be too novel, too confounding, to categorize? Questions like these, I knew, would fascinate Francis too.

Francis and I continued to correspond on various subjects, and I would always try to see him when I visited La Jolla. From 1997 to 2001 I was preoccupied with my memoir *Uncle Tungsten* and, less intensely, with matters of visual consciousness. I continued to see a stream of patients, however, and I often carried on a sort of mental dialogue with Francis whenever puzzling problems came up with regard to visual perception or awareness. What, I wondered, would Francis think of this — how would *he* attempt to explain it?

2

Francis's nonstop creativity — the incandescence that struck me when I first met him in 1986, allied to the way in which he always looked forward, saw years or decades of work ahead for himself and others — made one think of him as immortal. Indeed, well into his eighties, he continued to pour out a stream of brilliant and provocative papers, showing none of the fatigue, or fallings-off, or repetitions, of old age. It was in some ways a shock, therefore, early in 2003, to learn that he had run into serious medical problems. Perhaps this was in the back of my mind when I wrote to him in May of 2003 — but it was not the main reason why I wanted to make contact with him again.

I had found myself thinking of time the previous month — time and perception, time and consciousness, time and memory, time and music, time and animal movement. I had returned, in particular, to the question of whether the apparently continuous passage of time and movement given to us by our eyes was an illusion — whether in fact our visual experience consisted of a series of "moments" that were then welded together by some higher mechanism in the brain. I found myself referring again to the "cinematographic" sequences of stills described to me by migraine patients and which I myself had on occasion experienced. When I mentioned to Ralph Siegel that I had started writing on all this, he said, "You have to read Crick and Koch's latest paper — it came out just a couple of weeks ago in *Nature Neuroscience*. They propose in it that visual awareness really consists of a sequence of 'snapshots' — you are all thinking along the same lines."

I had already written a rough manuscript of an essay on time when I heard this, but now I read Crick and Koch's paper, "A Framework for Consciousness," with minute attention. It was this that stimulated me to write to both Francis and Christof (whom I had seen a few weeks earlier at Caltech), enclosing a draft of my article (entitled, at that point, "Perceptual Moments"). I threw in, for good measure, a copy of *Uncle Tungsten* and some other recent articles dealing with our favorite topic of vision. On June 5, 2003, Francis sent me a long letter, full of intellectual fire and cheerfulness, and with no hint of his illness.

> I have enjoyed reading the account of your early years. I also was helped by an uncle to do some elementary chemistry and glass blowing, though I never had your fascination with metals. Like you I was very impressed by the Periodic Table and by ideas about the structures of the atom. In fact, in my last year at Mill Hill [his school] I gave a talk on how the "Bohr atom," plus quantum mechanics, explained the Periodic Table, though I'm not sure how much of all that I really understood.

I was intrigued by Francis's reactions to *Uncle Tungsten,* and wrote back to ask him how much "continuity" he saw between that teenager at Mill Hill who talked about the Bohr atom, the physicist he had become, his later "double helix" self, and his present self. I quoted a letter that Freud had written to Karl Abraham in 1924 — Freud was sixty-eight then — in which he had said, "It is making severe demands on the unity of the personality to try and make me identify myself with the author of the paper on the spinal ganglia of the Petromyzon. Nevertheless I must be he."

In Crick's case, the seeming discontinuity was even greater, for Freud was a biologist from the beginning, even though his first interests were in the anatomy of primitive nervous systems. Francis, in contrast, had taken his undergraduate degree in physics, worked on magnetic mines during the war, and had gone on to do his doctoral work in physical chemistry. Only then, in his thirties — at an age when most researchers are already stuck in what they are doing — did he have a transformation, a "rebirth," as he was later to call it, and turn to biology. In his autobiography, *What Mad Pursuit,* he speaks of the difference between physics and biology:

> Natural selection almost always builds on what went before . . . It is the resulting complexity that makes biological organisms so hard to unscramble. The basic laws of physics can usually be expressed in simple mathematical form, and they are probably the same throughout the universe. The laws of biology, by contrast, are often only broad generalizations, since they describe rather elaborate (chemical) mechanisms that natural selection has evolved over millions of years . . . I myself knew very little biology, except in a rather general way, till I was over thirty . . . my first degree was in physics. It took me a little time to adjust to the rather different way of thinking necessary in biology. It was almost as if one had to be born again.

By the middle of 2003, Francis's illness was beginning to take its toll, and I began to receive letters from Christof Koch, who by that time was spending several days a week with him. Indeed, they had become so close, it seemed, that many of their thoughts were dialogic, emerging in the interaction between them, and what Christof wrote to me would condense the thoughts of them both. Many of his sentences would start, "Francis and I do have a few more questions about your own experience . . . Francis thinks this . . . Myself, I am not sure," and so on.

Crick, in response to my "Perceptual Moments" paper (a version of which was later published as "In the River of Consciousness"), quizzed me minutely on the rate of visual flicker experienced in migraine auras. It is only now, looking through our correspondence, that I realize these were matters we had discussed when we first met, in 1986. But this, apparently, we both forgot — certainly neither of us made any reference to our earlier letters. It is as if no resolution could be reached at that time, and both of us, in our different ways, shelved the matter, "forgot" it, and put it into our unconscious, where it would cook, incubate, for another fifteen years before reemerging. Francis and I both had a feeling of complementarity, I think, converging on a problem that had defeated us before and was now at least getting closer to an answer. My feeling of this was so intense in August of 2003 that I felt I had to make a visit, perhaps a final one, to see Francis in La Jolla.

I was in La Jolla for a week and made frequent visits to the Salk. There was a very sweet, noncompetitive air there (or so it seemed to me, as an outsider, in my brief visit), an atmosphere that had delighted Francis when he first went there in the mid-1970s and that had deepened, with his continued presence, ever since. And indeed, he was still, despite his age, a central figure there. Ralph pointed out his car to me, its license plate bearing just four letters: AT GC — the four nucleotides of DNA. And I was happy to see his tall figure one day going into the lab — still very erect, though walking slowly, perhaps painfully, with the aid of a cane.

I made an afternoon presentation one day, and just as I started, I saw Francis enter and take a seat quietly at the back. I noticed that his eyes were closed much of the time and thought he had fallen asleep — but when I finished, he asked a number of questions so piercing that I realized he had not missed a single word. His closed-eye appearance had deceived many visitors, I was told — but they might then find, to their cost, that these closed eyes veiled the sharpest attention, the clearest and deepest mind, they were ever likely to encounter.

On my last day in La Jolla, when Christof was visiting from Pasadena, we were all invited to come up to the Crick house for lunch with Francis and his wife, Odile. "Coming up" was no idle term — Ralph and I, driving, seemed to ascend continually, around one hairpin bend after another, until we reached the Crick house. It was a brilliantly sunny California day, and we all settled down at a table in front of the swimming pool (a pool where the water was violently blue — not, Francis said, because of the way the pool was painted, or the sky above it, but because the local water contained minute particles that, like dust, diffracted the light). Odile brought us various delicacies — salmon and shrimp, asparagus — and some special dishes that Francis, now on chemotherapy, was limited to eating. Though she did not join the conversation, I knew how closely Odile, an artist, followed all of Francis's work, if only from the fact that it was she who had drawn the double helix in the famous 1953 paper, and, fifty years later, a frozen runner, to illustrate the snapshot hypothesis in the 2003 paper that had so excited me.

Sitting next to Francis, I could see that his shaggy eyebrows had turned whiter and bushier than ever, and this deepened his sagelike and venerable appearance. But this was constantly belied by his twinkling eyes and mischievous sense of humor. Ralph was eager to present his latest work — a new form of optical imaging, which could show structures almost down to the cellular level in the living brain. It had never been possible to visualize brain structure and activity on this scale before, and it was on this "meso" scale that Crick and Edelman, whatever their previous disagreements, now located the functional structures of the brain.

Francis was very excited about Ralph's new technique and his pictures, but at the same time, he fired volleys of piercing questions at him, grilling him, interrogating, in a minute but also a kindly and constructive way.

Francis's closest relationship, other than with Odile, was clearly with Christof, his "son in science," and it was immensely moving to see how the two men, forty or more years apart in age and so different in temperament and background, had come to respect and love one another so deeply. (Christof is romantically, almost flamboyantly, physical, given to dangerous rock climbing and brilliantly colored shirts. Francis seemed almost ascetically cerebral, his thinking so unswayed by emotional biases and considerations that Christof occasionally compared him to Sherlock Holmes.) Francis spoke with great pride, a father's pride, of Christof's then-forthcoming book *The Quest for Consciousness,* and then of "all the work we will do after it is published." He outlined the dozens of investigations, years of work, that lay ahead — work especially stemming from the convergence of molecular biology with systems neuroscience. I wondered what Christof was thinking, Ralph too, for it was all too clear to us (and must have been clear to Francis too) that his health was declining fast and that he would never himself be able to see more than the beginning of that vast research scheme. Francis, I felt, had no fear of death, but his acceptance of it was tinged perhaps with sadness that he would not be alive to see the wonderful, almost unimaginable, scientific achievements of the twenty-first century. The central problem of consciousness and its neurobiological basis, he was convinced, would be fully understood, "solved," by 2030. "*You* will see it," he often said to Ralph, "and you may, Oliver, if you live to my age."

A few months later, in December 2003, I wrote to Francis again, enclosing a copy of the final proof of "In the River of Consciousness," emphasizing how much of it I owed to his paper in *Nature Neuroscience.* (I added that I had just been rereading *Life Itself* and found it even more wonderful on a second reading, especially since I had become deeply interested in the issues it raised. I enclosed a tiny article I had written for *Natural History* on the possibility of life elsewhere in the universe.)

In January 2004 I received the last letter I would get from Francis. He had read "In the River of Consciousness." "It reads very well," he wrote, "though I think a better title would have been 'Is Consciousness a River?' since the main thrust of the piece is that it may well not be." (I agreed with him.)

"Do come and have lunch again," his letter concluded.

DAVID SAMUELS

Buried Suns

FROM *Harper's Magazine*

THREE HUNDRED AND TWENTY FEET DEEP and nearly a quarter of a mile wide, the Sedan Crater still radiates energy from the explosion of a 104-kiloton nuclear bomb on July 6, 1962. One of ninety-eight nuclear devices exploded that year, the Sedan "shot" was detonated 635 feet beneath the surface of the Nevada desert. In less than thirteen seconds, the earth was emptied of 6.5 million cubic yards of sand and rock, some of which was lofted up into the atmosphere to return later as dust and rain, and the remainder of which was driven down into the earth or simply vaporized.

The Sedan Crater is perhaps the only place in the world where it is possible for a layman to comprehend the full force of a nuclear explosion. A wooden viewing platform stands above the hole like an abandoned lifeguard's tower, offering a quiet vantage from which to contemplate the implications of a giant radioactive sand trap that seems to absorb sound the way a black hole swallows light. The similarity between the cratered stretch of desert where I am standing and the moon was strong enough that the Apollo astronauts trained here and radioed back precise comparisons of the craters they found on the moon to the man-made craters of the test site.

The paucity of visitors to this silent and instructive place can be explained by the fact that the Sedan Crater is locked deep within the Nevada Test Site, a 1,375-mile preserve the size of Rhode Island, surrounded by fences and itself contained within an even larger secure archipelago of military bases where America's most advanced weapons are tested and stored. In the years since the

Trinity test at Alamogordo, New Mexico, there have been a total of 1,054 acknowledged American tests (24 of them conducted jointly with the British), 928 of which took place at the Nevada Test Site.[1]

This proving ground, on which the American nuclear telegraph fluttered seismic messages to the Soviet Union, was built by humble craftsmen who used recondite technologies to create a string of production plants and thousands of bombs that reshaped the geography of continents and redrew the boundaries between nations. The weapons that made American power possible were designed by teams of physicists and metallurgists and engineers at the Los Alamos National Laboratory in Los Alamos, New Mexico, Lawrence Livermore National Laboratory in Livermore, California, and the Sandia National Laboratories in Albuquerque. The secondary parts and components of the arsenal came from the Y-12 production plant located on the 35,000-acre Oak Ridge Reservation, about fifteen miles west of Knoxville, Tennessee. The plutonium and tritium that fueled the bomb came from reactors at the Savannah River Site in Aiken, South Carolina. The plutonium "pits" that powered the bombs were manufactured at Rocky Flats in Colorado. The finished weapons were assembled at the Pantex plant near Amarillo, Texas.

Eighty-five percent of the weapons in the current nuclear stockpile were designed by Los Alamos. The youngest warhead, the W-88, was introduced in 1991, and several of the key materials used in all nukes — including plutonium "pits" and tritium fuel — were last produced in the late 1980s.[2] By 2014 the W-88 and every other weapon in the stockpile will have reached the end of its intended life, and by then the United States might not have a single weapons designer left with test experience.

Responding to this possibility, Defense Secretary Donald Rumsfeld has recently appointed a Task Force on Nuclear Capabilities, whose job it will be to breathe new life into our capacity to design, produce, and test atomic weapons. But it may very well be that our chance to refurbish the nuclear program has already passed and that a resumption of testing is no longer practical or possible. As luck and time would have it, we still have with us people in a unique position to understand what a resumption of testing will mean and what its effectiveness might be, men and women who saw firsthand what their countrymen had for decades been conditioned to ignore and who spent the best years of their lives driving to work every day for the express purpose of setting off nuclear bombs.

The first successful containment of an underground nuclear explosion took place beneath the Nevada Test Site on September 19, 1957. When the test-site miners drilled back into the cavity after the blast, they first hit a layer of black frothy rock. Further inside they found gunk filled with little beads of radioactive glass. Frozen inside the glass were pockets of gas that contained a perfect microcosm of the universe in which they had been formed.

In the microseconds after the bomb was detonated, an egglike cavity began to bloom beneath the desert floor, quickly reaching temperatures of over 100 million degrees Celsius. A few seconds after the shot, the temperature inside the cavity had cooled to between 800 and 1,000 degrees centigrade, which is close to the melting point of the rock, allowing a mantle of glass to start to form inside the cavity. Over the next few minutes the temperature dropped again, and some of the glass fell to the bottom. By measuring the refraction of the light and the amount of water vapor dissolved in the glass, scientists found that they could estimate the size of the explosion with a very high degree of precision. Each kiloton of nuclear yield created approximately one kiloton of glass.

The last American nuclear test was conducted on September 23, 1992. A laconic press release issued by the Department of Energy on that date stated simply, "An underground nuclear test was conducted at 8:04 A.M. (PDT) today at the U.S. Department of Energy's Nevada Test Site." Code-named Divider, it was the sixth American nuclear test that year. Jim Magruder was running the Divider test from the Department of Energy's office in Las Vegas. A fit, taciturn man in his early sixties who looks like a retired astronaut and dresses like a real estate agent, he accompanies me on one of my first visits to the test site.

"I think once you shut a program down it's awfully hard to start it up," Magruder says in a flat Kentucky twang, as he gazes out at the crater through big, square, gold-rimmed glasses, "whether you have the personnel that used to do the work or not."

His companion, Larry Krenzien, dresses like an engineer, with a silver watch on his wrist and a practical, short-sleeved shirt. His voice is confident, easy, and deep, as befits a man who has worked on four hundred or five hundred nuclear tests during his lifetime, including Divider. He transferred permanently out to the test site in 1964, after working at Los Alamos and in the South Pacific during the atmospheric-testing program. Back when the air force would fly through the clouds after each test to collect debris, Krenzien was in charge of analyzing the radioactive samples from the clouds.

"I went out to Enewetak in '58 for two tests out there," he tells me. "Butternut, I think was one. Cactus was the other. They had steak every night, as many as you wanted. The beer was selling for five cents a can. It was pretty good living."

When I ask Krenzien about the glory years of exploding nuclear bombs at the Nevada Test Site, his face creases with pleasure. "I can remember that C-47 coming in with the device, and I would drive down in a regular flatbed truck and load it myself, get up there, put it in the building, lock the door, and go away," he recalls.

"The two-man rule wasn't very much respected," Magruder adds.

"Back in the early days, they shipped it complete," Krenzien explains. "It was one box, and that was it. At one time our warehouse man said he had more weapons than the USSR, and he probably was right."

Over the life of the American nuclear-design program, the scientists at Los Alamos and Livermore designed 71 different warheads for 116 nuclear weapons systems, at a total cost of nearly $800 billion. This year, the Department of Energy will spend $6.5 billion on nuclear weapons, and it plans to spend a total of $35 billion over the following four years, an amount that in real dollars equals what Ronald Reagan spent in eight years on nuclear weapons at the height of the Cold War.

"Do you think you saw the last American nuclear test?" I ask Krenzien.

"I don't know if I want to answer that," Krenzien says, turning to Magruder. "We have inexperience on the test site and inexperience in the labs."

"The last time we did a test there was nothing over fifteen stories in Las Vegas," he reflects. But with the revitalized plutonium-pit production scheduled for 2007, a multibillion-dollar tritium-production program funded by Congress, deliveries of refurbished nuclear weapons scheduled for 2006, and billions of dollars earmarked for the computers and visualization theaters at Livermore and Los Alamos, it is impossible to ignore what's in the offing. Both Stephen Younger, the former associate director of the weapons program at Los Alamos, and C. Paul Robinson, the current director of Sandia, have publicly advocated the development of a new generation of strategic nukes. As the Defense Department's Nuclear Posture Review explained to Congress at the beginning of 2002, "While the United States is making every effort to maintain the stockpile without additional nuclear testing, this may not be possible for the indefinite future."

Magruder, Krenzien, and I drive through a dry lake bed called Yucca Flat. The desert outside our tinted windows is dotted with juniper and yucca plants. The surrounding mountains are made of layers of welded volcanic ash. On our right are two skeletal frame structures and a lone wall, the remains of a remarkably detailed Japanese village built to scale in the middle of the desert as part of a 1962 experiment to determine the radiation levels experienced at Hiroshima.

We stop at a four-story square granary tower that stands over the 1,557-foot-deep hole dug for Icecap, the shot that was scheduled to go off a week after Bill Clinton put a "permanent" stop to the explosions in 1993. As we climb up into the tower we can see the 110-foot steel rack that would have housed the bomb and twelve experiments from the labs. The wind outside rattles the sections of the tower, making it necessary to shout to be heard.

"We were going to put in dry ice," Krenzien yells out, "and this was like a refrigerator to keep it cold. During the testing of some of the intercontinental ballistic missiles, they would encounter very cold temperatures in the outer atmospheres."

Behind the tower sits a trailer park where the instruments that were to collect data from the blast were kept. The Los Alamos trailers are white, and the Livermore trailers are silver. Inside each shock-mounted trailer were oscilloscopes, cameras, and other measuring devices.

"You could see a little ground motion on some of them," Krenzien shouts to me. "When Livermore was firing, and I didn't have to go to work, I would sit in my back yard and watch the swimming pool shake." He ducks his head into the collar of his jacket and shrugs his shoulders against the wind.

Jim Magruder shows me how each instrument was hooked up to the rack for Icecap. "It was always the policy that we would return to testing," he says, "but we always needed one more thing. We were all convinced it was a temporary shutdown." As Magruder explains it, Hazel O'Leary — the first secretary of energy during the Clinton administration — "had her test, the grandma test. Whether it made sense to her grandma. And soon it became pretty obvious that we weren't going to resume nuclear testing anytime soon."

We get back in the van and drive past the rows of subsidence craters and abandoned shot towers that dot the desert landscape. Some shots never "cratered," Krenzien tells me. Sometimes they'd shoot a test two years later and a nearby hole would suddenly collapse.

The flat desert basin we are passing through now is the back lot of the atomic age, the movie-ranch setting for most of the civil defense films of the 1950s that showed the effects of nuclear explosions on ordinary American towns and homes. For the Annie test on March 17, 1953, the men at the test site built a model of a small American town, with mannequins, school buses, cars, and fresh food flown in from San Francisco and Chicago. We stop the car at the remains of Priscilla, a 1957 test involving an underground parking garage filled with cars and a bank vault constructed by the Mosler Safe Company. The door of the safe was still intact. Sprouting from the ends of concrete columns are burnt ends of bent steel rebar that look like a book of matches someone lit all at once and then blew out. Ahead of me is a section of railway bridge built to European specifications that bears the faint but discernible imprint of a giant thumb. The shards of dull green glass at my feet are another indication that we are near ground zero of an atomic blast.

"You can tell the vintage of the tests by the letters," Krenzien points out once we are back in the car, heading for, or at least in search of, the final resting place of Divider. In his hand with the big silver wristwatch, he holds a map of more than nine hundred buried nuclear explosions.

"I think we make a right here," Krenzien says at the intersection of two rutted and weather-beaten roads with less than fifty yards separating the shot craters all around.

Magruder studies the map. "I still think that's it," he says.

"I remember it was a big, bare spot," Krenzien says. We get out of the van and study a cracked concrete pad from behind an uneven wire fence. It's the wrong hole; we keep driving. I ask Krenzien how he became the test director for the Los Alamos National Laboratory at the Nevada Test Site. He answers in a deep bass voice that has a little of the codeine-laced cough-syrup warmth of Rush Limbaugh.

"There was a tunnel shot coming up, and I remember I was in the dorm," he says. "And there was a knock on the door, and the guy who was the test director on this tunnel shot stuck his head in and said, 'I'm not going to do this job anymore. You're it. Goodbye.' He was only the test director for one shot. So I inherited it, really."

"There sure aren't many cables. It looks like an older one," says Magruder, buried in the map.

"That would not be it then," Krenzien responds. The man who helped run the last American nuclear test directs our driver to take the next right, then a left. As we pull down a rutted road, Krenzien sticks his head out the open window and we stop.

We step out of the van to see big strands of cable descending into an enormous, bowl-shaped depression in the sand surrounded by a plain wire fence.

"The trailers would have been sitting here," Magruder says, pointing to a spot beyond the sunken ground. "And these cables went to each individual trailer. And this line of posts down the center would have been the cables connected to the firing device."

"Looks like there were maybe seven trailers on it," Krenzien says. There is a dark patch in the middle of the crater and a ring of light sand around the edge. A nuclear blast, Krenzien explains, affects the soil "so that beautiful flowers come up in the spring."

I ask him again if he thinks that he shot the last American nuclear test. He considers the question for a moment, looking out over the pitted landscape he helped to create, and then he says, "I don't know that I can see us starting a test series ever again, as such."

The attempt to reconstitute a nuclear complex the nation spent a decade dismantling is not simply a matter of spending more money and building more bombs. It involves the reclamation of a craft that is already more than half forgotten. Bred of science and military might, caution and awesome destructive power, extraordinary technical ingenuity and hundreds of billions of dollars, the American nuclear testing program was also undeniably the work of thousands of individuals whose skills were honed over decades of secret work in the weapons complex. The men who shaped the Nevada Test Site, who built imaginary cities and blew them up, who drilled holes in the mesas and the volcanic rock, who readied the devices and lowered them into the holes, who caused the explosions that announced America's resolve to its enemies, are mostly dead or retired. In the precision of their knowledge, in the plainness of their speech, the men of the test site are the voice of the effort that made the America we inhabit today. The mementos that hang on the walls of their modest homes in and around Las Vegas tell the story of how we faced the infinite, without recourse to the affirmation or denial of God, and won at the game of nuclear poker we played with the Russians.

William Gus Flangas

Bill Flangas was the man in charge of drilling holes in the mesa. On the wall of his suburban ranch house is a photograph of himself as a wiry, tough-looking little sailor in 1945, on a bridge in the middle of Tokyo near Emperor Hirohito's palace. There are pictures as well of four generations of Flangas men, who all appear to be cut from the same cloth, with extra fabric allotted for ears: there is Gus William Flangas, who begat William Gus Flangas, who begat Gus William, who begat William Gus.

The current William Gus Flangas was born in June 1927 in Ely, Nevada, a copper town. Ely was one of the main towns in White Pine County, which also produced silver, gold, and molybdenum. The mining was done mainly in Ruth. When Flangas came home from Japan he became an engineer and went to work in the copper mines. As Flangas tells it, a man named Reynolds, who ran a mining concern that had been awarded the mining contract for the Nevada Test Site, "was hobnobbing with some Rotarians," looking for some mining engineers to take the nuclear program underground. "And that's where my name popped up."

With the support of his new boss, Reynolds, Flangas helped hire on the first real miners at the test site from some local mines that were about to close down.

"Yeah, I was very fond of him," Flangas says of Reynolds, who was over six feet tall and had started his company in El Paso in 1932. "He felt that if the customer was heading into a direction that he ought not to be, that you should have the courage to tell him that maybe this isn't right."

One of Flangas's first jobs at the test site was to take a team of miners to the mesa and drill back toward Rainier, the first underground nuclear explosion. No one knew what to expect. They drilled back hundreds of feet through the rock until they encountered a layer of glass. "Black, red, purple, all kinds of colors," he remembers. "When we entered the cavity, there was just about an inch or more of plaster there, and it was very, very radioactive. And so we plastered that with lead." Fifteen months after the Rainier shot, the ground temperatures inside the cavity were upward of 160 degrees, and the ground was filled with radioactive water. The miners used dynamite to clear their way through hot, wet ground filled with steam from a buried nuclear explosion. Accompanying Flangas and the miners was a Livermore physicist named Gary Higgens.

"Oh, yeah, he was raised on a farm in Minnesota and he went all the way to become a Ph.D. physicist," Flangas remembers. Higgens got along well with the miners. "They trusted him and he trusted them, and lo and behold that was one of the initial major breakthroughs in the underground testing."

When a group including Flangas became contaminated with tritium in 1959, it was Higgens who helped come up with the idea of isolating the contaminated men in a room and having them effect a full fluid exchange in their bodies by drinking cases of beer.

"You know, if you want the maximum fluid in and the lowest retention — in other words, you want to be pouring it in on one end and a constant stream out the other end . . . ," Flangas explains. "I had one Latter-day Saints guy there, my hoist man, and he refused. Which was okay, he was in his sixties then, and so he gave us a reference point. We just sat there, and we drank beer, and we drank beer, and we drank beer."

Fred Huckabee

Fred Huckabee has a mournful face like Lyndon B. Johnson's, a fellow product of the Texas hill country. The license plates on the

Buick Le Sabre in his driveway read HUCK. He smiles when I ask him about the excitement of working with nuclear explosives. Hanging on the wall near the door is a medal graced by the American and Soviet flags and dated 1988. On the other side of the medal is a Russian inscription that reads IN MEMORY OF THE VISIT over the unmistakable image of a mushroom cloud. He keeps his house clean, with white place mats on the table at which he sits, and sips water from a thirty-four-ounce plastic convenience-store cup and watches golf on his big silver RCA color TV.

Huck, as he is universally known, was born on a ranch in Bramfield, Texas. After high school he went to work for Magnolia Petroleum and roughnecked on drill rigs in Midland and Odessa and up through New Mexico. He graduated to become a driller with the Great Western Drilling company, then a rig superintendent, otherwise known as a tool pusher, a job that paid well enough for him to marry a girl from Sonora and buy a twelve-foot-wide trailer. When the Russians broke the moratorium on testing in 1961, John F. Kennedy ordered the Nevada Test Site to swing back into action, and Huck and his two closest friends from the oil field headed out to Nevada with their rig.

"You could look off the mesa after dark," he recalls, "and it'd look like a city down there of vehicles running around, you know, car lights running around, drill rigs lit up and drilling holes, and it just looked like a city down there."

Huck's memories of the site range from drilling back toward exploded bombs to attempting to use nuclear explosives for commercial purposes in the mining and drilling industries.[3] But the most memorable adventure of all, as far as Huck is concerned, was the trip to Russia in 1988 and the visit to the Semipalatinsk range, the top-secret nuclear installation in Kazakhstan that was the Soviet equivalent of the Nevada Test Site.

"They fed us a lot of tongue and mare's milk," Huck says. "That's a mare of a horse, you know," he explains. "It had buttermilk, and I liked that buttermilk, boy, but a bunch of 'em wouldn't even touch it. They didn't have salads or anything, and they'd feed us egg sometime, and with a tongue, with our breakfast. A lot of guys didn't like their food, but hell, I ate everything they had." In the evenings the Americans watched movies from home and made popcorn, which they shared with their Russian translators.

What Fred Huckabee also saw was that the foe America had been fighting for forty years was using technology that was twenty and thirty years out of date. "They set up an army tent with little fireboxes inside, for heat, and it was a double-doored army tent, thirty by thirty or something like that, and instead of having projectors to give their presentations of how they tested and everything, they'd put a piece of paper up on a wire." When the Russians came to the Nevada Test Site, they were equally amazed by what they saw. Their American hosts took them to see Siegfried and Roy and then to a Smith's Food King off U.S. 95. The Russian delegates all bought children's pacifiers. One of the colonels bought a Timex watch, and Gorbachev's personal interpreter bought a three-pound tin of Folger's coffee for his boss.

J.L. Smith

J.L. Smith was the man responsible for preparing the shack from which the final shot was fired. In his late sixties, he lives in a neighborhood of low-built ranch houses near the Desert Mesa power plant, just around the corner from the Department of Energy in Las Vegas. Outside his house a gray Cadillac is parked on the curb next to a desert-baked, baby blue Chrysler white-top pickup. On the day I meet him, Smith is wearing a dirty Dallas Cowboys hat with a star on it, a white walrus mustache, and a big, warm smile.

"J.L.," he says, sticking out an ample palm. "That's initials only." He welcomes me into his low-ceilinged home, where pictures of angels and of Jesus and his disciples occupy the wall above the couch. The wall near the window is filled by a fifty-six-inch Magnavox TV. Born in Mississippi in 1935, J.L. was raised on an eighty-acre farm and graduated from Betty May Jacks High School in Morton. His stepfather, whose name was J. D. Heald, liked to dress up like a cowboy and come out to Las Vegas every year for the rodeo. J.L. joined his brother in Las Vegas in 1955 and got a job building the first housing project in town. He joined labor local 872, which sent him over to the test site. As he sits beneath a picture of Jesus and tells his story, tears stream down his cheeks, though the expression on his face never changes. Allergies, he explains.

"Well, we would dig ditches and put in piping, build houses and blow 'em up," he says, when I ask him what he did at the test site.

"We built a whole city one time, a whole city that had, like, the train station, the railroad, and there'd be railroad tracks and houses with people — dummies — sitting at the table like they were eating supper or something," he recalls. "They had fruit and stuff on the shelves, just like it was at your own home in the kitchen." The bric-a-brac in his living room includes old pictures of tumbledown cabins, a hand-painted statue of James Brown, a singing rabbit with a pair of antlers, and a fish that sings "Don't Worry, Be Happy." I am curious to learn more about the shack where the atom bombs were armed and fired.

The later shacks were made of metal, but the first ones were made of wood. They would be brought out to the hole on a flatbed truck. "Yeah, just a little shack," he says. "One room, and sometimes they had partitions in there," he explains. Once the marriage of the bomb and the rack was made, the scientists would open up the canister and arm the device.

I ask if he ever saw what was inside the bomb itself.

"Mm-hmm," he answers. "It looks like half of an egg, like a little egg-shaped thing, about that big around."

"Like the size of a softball?" I prompt him.

"Softball, about the size of a softball, about the size of half of a softball, something like that, like you cut the softball in half. It looked something like a foam, Styrofoam or something like that."[4]

"So there was something inside that made a lot of —"

"Made a lot of boom."

There were also hundreds of wires, he remembers, that suggested a wasp's nest. "They were small wires, you know. And it takes an hour, hour and a half, to get them all hooked up." As the men armed the bomb, they talked about neutral subjects. "They talked about what happened last night at the cafeteria and stuff like that. 'I saw a show last night,' that they had had a good shrimp dinner or lobster dinner or something like that, you know. Just usual, comfortable conversation."

Roma B. Washington

Crane operator Roma B. Washington, who lowered the armed bombs down the holes at the test site, was in fact waiting for a bomb to be delivered to his crane when the American nuclear testing program ended in 1993. He is a choleric little man built like a cannonball. "We was delivering the last day in Greenwater. I was the man," he remembers. "I was up on the last one. I got all pumped up and everything. Then two hours later they told us, 'Hey, shut it down. Lock the building up, you got thirty minutes to get all your tools out,' and all that kind of stuff. I mean, it was trippy, like *The Twilight Zone* or something."

Washington's nut-brown skin flushes red beneath a gray wife-beater T-shirt. "I'd come home to a wife and everything, and I'd come home to a family, and I'd never mention nothing about I just set a bomb in a hole." He sighs heavily, putting his hand on his knee. "Because, you know, somehow or another, it's just a unrewarding job." He came to Las Vegas from Louisiana, got a job as a sauté chef, became a crane operator as part of an equal-opportunity effort in the unions, and was sent out to the test site. His specialty was putting the bomb to the rack, otherwise known as making the marriage.

"You swing the rack over, right over top of it, and then just lower it down. Then they marry it, with bolts and stuff, you know what I mean. Pick it up, take it to the tower, then they do that thing again, whatever they doing," he says, like an embarrassed dad laying out the specifics of a sex act to an underage son. "After they done what they do, and all the wires and all that," he continues, "you bring it out of the building, which was a very, a very technical move, and then you start lowering it in." Sometimes the process of lowering the rack down into the hole took three days. Sometimes it took eight. Washington's knee jumps up and down. "You lowered it down to a point, then you stopped and they did their little deal, then you'd slowly lower it down to another point, and they'd say, 'Stop.'"

I ask him how he shielded his mind from the knowledge of what he had at the end of his crane. He stares at me for a while before he answers.

"Some people, when they were to do that, it kind of worked on them. It kind of ate them up, you know? You could see them slowly deteriorate and all that. But it was all a mental thing with me, and I'm that way to this very day. My crane was kind of like a wife to me. I trusted her. I named every crane I ran."

"What were some of their names?" I ask.

"Well, that one was Gargantuous." He laughs. "That was Big Foot, and that was Lucille, Spirit of '76, Bertha, Piece of Shit, Miracle of 17 Hill."

"What was that one named for?"

"Well, that's another story," he says. "I had nothing to do with it."

The night before he lowered a bomb down the hole, he says, he always slept like a baby. I ask him about his feelings afterward.

"Well, what are you trying to say?" he asks, as his leg starts jumping again. "Do I go home and have a drink? Do I get drunk? Do I smoke a joint? Is that what you saying? You ain't as slick as you think you are," Washington declares. "Well, what if I said I came home and had sex with the old lady?"

I try to get back on his good side by suggesting that a little sex and whiskey seems like a reasonable enough response to the stress of operating a crane with a nuclear bomb attached to it.

"I'm not going to tell you that I don't come home and have a beer, or I don't come home and have a shot of brandy, you know, after a day. I mean, I'm not what you'd call a stone drinker, but I like to have a sip every now and then."

I ask him again what it was like to help win the Cold War.

"I felt good about being a part of that history, the things that advanced America, advanced the world, you know what I mean? Now, as far as being something that's about blowing up a bunch of people, no, I wasn't proud of that, no."

Sitting with these men during the afternoons and the evenings, I found it impossible not to feel that their work at the test site had shifted their angle on reality in a subtle but permanent way that separated them from other patriotic retirees in the Vegas area. Their literal, factual accounts of stemming procedures and post-shot operations were often delivered with a deadpan humor that suggested a shared cosmic joke burbling just below the surface of whatever question was being asked or answered. I suspect that having spent the better part of their lives working in a secret world they could not share with their wives and neighbors, a world pervaded by the daily aftertaste of inconceivable destructive power, the men of the test site had learned that the world was cracked. In order to see what they had seen, or as much of it as would be shown me, I made arrangements to go down into the tunnels one afternoon and to get as close as I will likely ever come to the actuality of a nuclear explosion.

Just across the Mercury Highway from Divider is a green shot tower for Unicorn, a "subcritical" test that will happen sometime this year. A few hundred yards away from the Unicorn tower is the grave site of Ledoux, an underground nuclear test from 1990. Ledoux is now embedded in the U1a Complex, a tunnel system dug in the 1980s and 1990s to serve as a reusable home beneath the desert for low-yield nuclear tests.

I am met at the top of the shaft by Rafi, a cheery bearded man who works beneath the desert as the test-group director for Los Alamos. His full name is Ghazar Papazian, and he was born in Egypt, he tells me, as I lace on a pair of miner's boots next to a sticker that reads, "If you ain't a miner, you ain't shit." When I ask him what America learned from setting off more than a thousand nuclear explosions on, or under, our own soil, he gives me a scientist's answer:

"What's odd about that data set is that for every nuclear test, the codes that run that nation, if you will, are calibrated to that specific test. So for system A, if it had six different tests, the density would change for each of those tests. Because to make the code work and give you the right answer, you had to turn knobs." We pass through a security gate and get ready to descend into the hole. "What we're trying to do with subcrits is to eliminate the knob turning and give the folks who do code development a single data point as far as what the density of plutonium is at a specific pressure."

Eventually, and with much patience from Rafi, two extraordinary ideas enter my brain: (a) that the entirety of 1,054 American nuclear tests has netted us less than a single second's worth of usable data, and (b) that we still know far less than we need to about the properties of plutonium.

The elevator door closes with an undersea clang. A foghorn sounds, and we descend 963 feet to the bottom of the shaft. When the grinding noise of the elevator ceases, a gate opens onto an anteroom with a framed quotation from President Woodrow Wilson. "To the miner," it reads, "let me say that he stands where the farmer does; the work of the world waits on him."

Nearly one thousand feet below the desert, smoking is definitely allowed. Near the miners' battered lockers is a host of big old hotel ashtrays.

"We overventilate the shaft," Rafi explains. Beyond him is a long gray hallway that looks like a bomb-shelter tunnel from the 1960s.

Rafi points toward the side of the shaft, where a sign reads DANGER: HIGH CONTAMINATION.

"That's the entrance into Ledoux," he says.

"What was Ledoux designed to prove?" I ask.

"It was part of Star Wars," Rafi answers shortly.

I stand for a moment in silence in the tunnel and try to sense the energy of a dead sun.

"It doesn't compute in your mind," Rafi says, "the amount of power that's in your hand. You can hold it."

"You've held nuclear devices in your *hands?*" I ask.

"Sure. I've held test devices as we've put them into the experiment. I've held pits of TA-55 when they're produced."[5]

We pass by Kismet, a radioactive room sealed behind a steel door like an inadmissible thought. "It was basically a five-gallon drum of explosives and a commercial X-ray head," Rafi explains. In other words, Kismet was a dirty bomb of the type that any terrorist might explode in an American city. I peer through protective doors into a shattered room illuminated by a bare light bulb. Rafi sometimes sends a mining-rescue team in to change the light bulb.

I ask him whether Kismet went smoothly. He shakes his head. "A miserable test," he says. "We lost all that data. People worked in bunkers. They worked in labs. They picked up some really nasty habits. The lesson learned was that if you're going to have a nuclear-readiness program based on an experimental program based on the Nevada Test Site, you can't just work in bunkers."

Right now, Rafi explains, one of the main focuses of the atomic testing program is to examine the old plutonium pits, manufactured at the now-closed Rocky Flats plant outside of Denver, that make up the bulk of our nuclear arsenal and even to replace some of those pits with new ones made at Los Alamos. "The question is, how does it age? If you were to go and find a designer from the past, and said, 'Okay, you designed system XYZ, tell me what you thought the life span of that device was,' he'd tell you, 'Fifteen to twenty years.' We have systems in the stockpile today that are fifty years old. How do you gain confidence that a fifty-year-old pit will function as originally designed without having to test it?"

In the background a voice echoes through the tunnels from a distant loudspeaker — "Five, four, three, two, one. One, two, three, four, five" — and for a split second I wonder if a bomb isn't about to go off.

Rafi stays calm. We walk through the tunnels until we come upon a cluster of miners and other refugees from the glory days of the test site.

"This is Chuck Eaton. He's the miner in charge," Rafi says, introducing me to a burly man with an incredibly deep bass voice. I ask Chuck how long he's been down here. "I've been underground since '72," he says. "I did construction mining, development mines, sinking shafts, until I came here." The old days, he tells me, bear very little resemblance to what I am seeing now.

"There were tunnels, and pipe shots, and it was a whole different ball game than this." When I ask him what impressed him the most about his work, he shakes his head, smiles, and answers without hesitation: "Well, setting off a nuclear bomb."

Sitting at the counter in a darkened sports bar, safe from the flat Las Vegas sunlight, Nick Aquilina takes another sip of his Diet Coke and says something nice to the waitress. This is a place he likes to come to in the afternoons, he says. A handsome man in his sixties, he has the face of a wise old janitor at a local high school who lets the kids sleep in his office and finds them jobs raking leaves. He wears his white sideburns neatly trimmed, and his thick salt-and-pepper hair is fixed with pomade. The gold watch on his wrist attests to decades of faithful service to the testing program, and the golf ball embroidered on his chest testifies to his current love of leisure. "There's never a bad day at the test site," he says breezily.

On September 11, 2001, Nick Aquilina was with Troy Wade in Washington. The two men had been called to Washington by the agency in charge of the country's nuclear weapons, he tells me. They were meeting at the L'Enfant Plaza Hotel, which is right up the street from the Department of Energy, where a number of veterans of the testing program were discussing, in very general terms, how the country's nuclear bureaucracy should be organized and what it might take to restart the country's nuclear testing program, when an airplane hit the Pentagon, less than a mile away.

"Bob Kuckuck was there. Don Pearman was there, and Troy and I," he says, naming several luminaries of the test program and the labs. When the men at the meeting stepped outside, they could see smoke coming from the Pentagon.

That night three of the old-timers — Aquilina, Wade, and Kuckuck — met for dinner at an Italian restaurant. Together the three men had helped to develop and test nearly every weapon in the American nuclear arsenal.

"I think we were mostly talking about what was going to happen the next day," Aquilina remembers. "We were concerned. Was this the first in a series of things? Was something going to happen the next day? And how can we respond to something like this? Who do you respond to?"

After employing 120,000 men to drill about a million feet worth of holes in the desert, or roughly the distance from New York to Washington, and to blow up more than a thousand nuclear bombs, the test site and its work were over and done with. The secret that bound the men of the test site together was the firm knowledge that neither the United States nor the Soviet Union had actually intended to use nuclear weapons. The warheads were poker chips, and the test site was the table on which America bluffed the Soviets into folding. But the Islamic zealots who attacked the World Trade Center and the Pentagon with bombs fashioned out of passenger planes did not have a military complex that could be bluffed with nuclear weapons. The familiar strategic equation of American global supremacy was reversed: no longer the masters of the plutonium age, we were now its potential victims, living in fear of the day the terrorists gain possession of an atomic weapon detached from one of the dissolving state structures left over from the Cold War.

What the men who ran the test site could not have foreseen on the evening of September 11 was that the absence of a viable strategy against this new opponent might prompt a return to the table with the hopes of playing the same hand twice.

The element plutonium, a heavy metal, was first formed in the blowoff from supernovae that exploded after the big bang and made all the elements in our sun and its planets. Plutonium was discovered by mankind on February 23, 1941, when Glenn Seaborg, a young chemist with a Ph.D. from Berkeley, isolated a minute quantity of what he and his colleagues at first called element 94.[6] The first one-gram shipment of plutonium arrived at Los Alamos in February of 1944. Much of the energy at Los Alamos National Laboratory since that date has been spent in the attempt to engineer plutonium into a stable form. "After more than fifty years of plutonium research at Los Alamos, we might be expected to understand the strange properties of this metal," wrote George Chapline and James L. Smith in "An Update: Plutonium and Quantum Criticality," one of several dozen recent publications on the subject by Los Alamos scientists. "Instead, we are still stumped."

Nearly twice as heavy as gold, plutonium is silvery, radioactive, and toxic. The pure metal first delivered to Los Alamos showed wildly differing densities, and the molten state was so reactive that it corroded nearly every container it encountered. Happier as a liquid than as a solid, plutonium has seven distinct crystallographic phases and the highly democratic ability to combine with nearly every other element on the periodic table. It can change its density by 25 percent in response to minor changes in its environment. It can be as brittle as glass or as malleable as aluminum. Chips of plutonium can spontaneously ignite at temperatures of 150 to 200 degrees Celsius. When crushed by an explosive charge, plutonium's density increases, which decreases the distance between its nuclei, eventually causing the metal to release large amounts of energy — enough to vaporize a city.

Jay Norman, a former director of the weapons testing program at Los Alamos and the current deputy manager of testing and operations for the National Nuclear Security Administration, meets with me one afternoon in Las Vegas to explain how a nuclear bomb is designed and tested. A pleasant-looking, soft-spoken man with sharp features, clear, rimless eyeglasses, and hair that stands on end in seeming tribute to popular conceptions of the mad scientist, Norman wears a tropical white linen jacket and a grape-colored cotton shirt that balloons gently over his belly.

Although Norman is not allowed to talk about the weapons he makes, it is not all that difficult to explain the way a nuclear bomb works. A modern thermonuclear weapon consists of three operative elements: a primary, a secondary, and a radiation case that encloses the parts. The plutonium pit is surrounded by high explosives. This is the primary.

Detonating the mantle of high explosives surrounding the plutonium squeezes the primary into a supercritical configuration. The energy from the primary is then trapped and forced to surround the secondary part of the device, which is a highly combustible radioactive substance such as uranium. Enormous pressures are created, and the secondary implodes, releasing nuclear yield.

The goal of most of the nuclear-weapons designs of the last four decades or so consists in getting bigger yields with less weight by refining existing designs and materials. As a result, the nuclear package or "gadget" that weighed 10,000 pounds at the Trinity test now weighs 250 pounds and can fit inside a suitcase.

In the early phases of a design for a new weapon, Norman explains, Los Alamos and Livermore would compete to develop a winning design that would meet the specifications handed down by the Defense Department and the Department of Energy. "Livermore might say, 'Well, I think it ought to look like this and this,' and Los Alamos would say, 'Well, let's do this and this,' and each lab would go out and test their proposed configuration, and each lab would learn something." But competition has hardly led to variety in our nuclear arsenal. "Without trivializing it or making it too mundane," Norman says, "there's a lot of similarities in the stockpile, like a Ford sedan, and a Chevy sedan, a —"

"Chrysler?"

"There you go, a Chrysler," Norman says.

"No Lexus or Mercedes?" I ask. Norman cocks his head and gives me a strange look.

"A Lexus is just a more expensive car," he says tentatively.

"So there's plenty of Lexuses," I conclude. Norman nods.[7]

"I think you know that in 1989 the last order for a new nuclear weapon was canceled and that the last nuclear test was Divider, which was three years later, in 1992," he explains. "But there were always questions about would this be better than that, and would this, and so . . ." His voice trails off.

When I suggest that the resumption of the American nuclear testing program sometime over the next five years or so seems inevitable, if only to ensure the reliability of the current stockpile, he says, "See, my basic fear of stewardship, of [proceeding] without nuclear testing, is the following: it's not that confidence will degrade in the stockpile. It's that confidence will absolutely soar. You can calculate yourself off into this computer-generated three-D holodeck where everything is perfect, when the things you are looking at might not actually work."

The Atomic Test Site Museum, a mile and a half from the Vegas Strip, is today the center of a shadow world populated by the men who once lived with the madman's knowledge of total destruction out in the desert. The museum's founding spirit is Troy Wade, who was with Nick Aquilina in Washington on September 11. Hawkishly handsome, at seventy-one Wade dresses like a jazz piano player, in a black turtleneck and elegant gold chains that dangle from his wrists. He grew up in Cripple Creek, Colorado. His mother went to Juilliard and taught high school. His father played C-melody saxophone in a band. Wade began his career at the test-site mining tunnels, then ran Livermore's explosives assembly facility, and finally wound up as an assistant secretary of energy in the second Reagan administration, responsible for overseeing the secret world of the bomb.

The walls of Wade's office in the Frank H. Rogers Science and Technology Building are bare except for an aerial view of the Pentagon and an aerial view of the test site. "Are you familiar with the Sedan Crater?" he asks me, pointing up to a large crater in the middle of the upper quadrant of the test-site photograph. "Well, the very first nuclear weapon that was totally my responsibility — that I put together with these two hands — was Sedan, which is not a bad way to start in the business." After some pleasant conversation, he graciously agrees to walk me through the half-finished halls of the museum.

"So, when we open, this is the Wackenhut Guard Station," he says. "You're going to buy your ticket here. You'll get a ticket that looks like this, looks like a badge. It's going to be about eight or ten bucks." I am curious to know why Wackenhut is being honored in this way. "Wackenhut's first major contract was at the Nevada Test Site in the mid-sixties," Wade explains, "and two years ago he sold the company to a Danish corporation for $570 million. He gave me half a million bucks for the museum."

"The message we have to tell people here," he explains, "is why in the world did the United States get into this business? So in ninety seconds you're going to learn about what was happening in Nazi Germany, a little bit about what was going on in the Soviet Union, the Einstein letter to Roosevelt that sort of said, 'If we don't do this, somebody else is going to.'" The next exhibit concerns the history of atmospheric testing at the Nevada Test Site, which began at 5:45 A.M. on January 27, 1951, when a one-kiloton bomb named Able was detonated above Frenchman Flat.

"That is the entrance to the underground test area," he says, as we pass into the next exhibit hall. "This is what it's going to look like — it's all covered in rock." He pauses by a one-fifth scale model of a bomb rack like the one I saw out at the test site. The next room contains a big grain silo, on the side of which is painted the word SIOUX with an arrow through it. "There will be a little theater in the round in here that will tell the story of the experimental farm, including the fistulated steers, where they were measuring the uptake of radionuclides into the food."

The museum will also house the McGuffin collection of Piute Shoshone Indian artifacts collected on the test site by Don McGuffin, who gave it to Charlie Neeld at Livermore, who, when he was dying of cancer, gave it to Wade. After some negotiations, the tribal elders of the Shoshone Nation agreed to keep the collection together and allow it to be displayed at the museum.

We discuss the fact that all these hundreds of tests out in the desert gave us less than a minute of data, and his eyes glimmer. "It's actually less than a second," he tells me. "It's a fraction of a second. It was millionths of a second." I ponder the smallness of the knowledge gained against the size of the mushroom clouds this man witnessed.

"Let me tell you," he says, "the specter of seeing this mushroom cloud . . . You can actually see the shock wave coming across the desert at you, and when it hits you, you feel the thermal effects and you feel the shock, and it'll knock you right flat on your butt. I became a better American and a better patriot and a better God-fearing man because I did see them."

During the day the ghosts of the American nuclear testing program tend to stop by the museum. They drop off stacks of faded company newsletters or shoeboxes full of patches and certificates attesting to the role they played in making the desert bloom. One such ghost is seventy-four-year-old Ernie Williams, a white-haired, blue-eyed Nebraskan who looks a little like Johnny Carson. He is dressed like a prosperous alfalfa farmer, in a white shirt with silver snaps and a silver key ring fastened to his belt that holds the keys to the atomic kingdom. In his spare time, he works here at the museum as a security guard. By his own estimation, he saw about eighty atmospheric tests before the program went underground. He enlisted in the air force in 1951, after graduating from high school. Assigned to Lacklund Air Force Base, he scored high on an aptitude test and was sent to school to study the assembly and disassembly of nuclear weapons. After leaving the air force, he took a job with the Atomic Energy Commission and was sent to Enewetak Atoll in the Marshall Islands, where he soon became acquainted with Edward Teller, the father of the hydrogen bomb.

"Dr. Teller was the type of an individual, if you were the janitor, you thought you were talking to another janitor," he says. "He never came across to you as 'I am the father, I am the physicist.'" While waiting for the bomb to go off, the men played rounds of penny-ante poker, which Edward Teller inevitably won. "You'd just say, 'Oh, I'll throw my cards in,' and then he'd turn his cards up and he wouldn't have anything," he recalls. "But that's the nature of playing cards."

It was only after Williams went to work at the Nevada Test Site that he saw a vision of the apocalypse he had been assembling with his hands. In the desert there were rows of trenches, with fifteen to twenty soldiers in each trench waiting for the bomb to go off at 4:30 in the morning. The trenches were about five feet deep and eighteen to twenty inches wide. The desert was pitch black. Jeeps and trucks with army markings were parked at varying distances from the shot tower, and dogs and pigs were staked at varying distances from ground zero.

"You'd hear the countdown: 'One minute. Put your goggles on.' Everybody got their goggles on, and then you would hear, 'Fifteen seconds.' And the next one would be 'Ten seconds,' and then you would hear, 'Nine, eight, seven, six, five, four, three, two, one, zero.' And when zero hit, you know, a bright light, extreme bright light — even with the 4.2-density goggles on, it's extremely white. The mountain range around when it goes off is just like daylight, just like the sun's shining."

The soldiers were understandably jumpy under these conditions, and Williams recalls that officers were present "to direct and say, 'Look, you know, it's okay. It's not a problem.'"[8] Once the shock wave had passed, the troops were marched out of the trenches over the radioactive ground toward ground zero.

While serving as the proving ground for nuclear war, the Nevada Test Site was also its sound stage, the source of photographs of weird protoplasmic blobs erupting from the desert, those civil defense films depicting the effects of a nuclear blast on the average American family, and many of the spectacular atmospheric blasts that are now shown every fifteen minutes in the Atomic Test Site Museum's surround-sound theater.

The source of many of these images, and the technology that captured them, is a company called EG&G, the MGM of the Cold War. The partnership that would become EG&G began in 1931 with the invention of the stroboscopic flash, the first timed flash unit for photographers, by an MIT instructor named Harold "Doc" Edgerton and Kenneth Germeshausen, his assistant. Their flash-timing system would prove to be a key element in the design of the electrical detonating system for the bomb.

One afternoon while I am watching films of mushroom clouds at the museum, I meet Pete Zavattaro, the former head of EG&G. In his white polo shirt with navy piping and tan chinos, his white beard and white mustache, he looks like a former record-company chairman who has turned late in life to charitable works. As we speak, Zavattaro flips through a book of EG&G photographs of the atmospheric shots taken with RAPATRONIC cameras that eventually achieved exposure times of less than a millionth of a second. By the time of the Divider test, Zavattaro says, the dream of the photographers was almost in reach.

"To be able to actually see this radiation flow in three dimensions was the ultimate goal," he says, a little sadly, "and we were getting closer and closer."

I sympathize with Zavattaro's longing to capture the ultimate knowledge that lies at the core of the nuclear explosions he recorded over decades at the test site. Nanosecond by nanosecond, expanding in real time, a picture of creation might yet emerge. It is hard not to see the quest for the Ur moment as a secular equivalent to the yearning of the prophets to see the face of God. What makes Zavattaro and his fellow test-site veterans sad is the knowledge that the asymptotic approach to a final understanding of the atomic bomb has ended. There are few poker tables on earth where the same bluff will work twice. God has disappeared forever from the desert.

Toward the end of my stay at the museum, I spend some time down by the guard station with Layton O'Neill, a little man with happy, nutty eyes and thinning hair. As far as he is concerned, fears of radiation and atom bombs are so much nonsense. He is seventy-seven years old and has the energy of a mountain goat. His security-guard badge is fastened to his blue shirt, which he wears with a black string tie and a decorative clasp in the shape of the state of Nevada. When I ask him for the secret of his bountiful energy, he credits the small doses of radiation he absorbed during his days at the test site.

He believes that "small amounts of radiation are beneficial to humans," referring me to the researches of a man he calls "Dr. Lucky."

As much as anyone, Layton O'Neill is responsible for the effort to preserve the history of the Nevada Test Site. He tells me he comes from a nuclear family. "I have a brother who was making plutonium up at Hanford," he explains, in his excitable way. "I had another brother in California building guidance systems for submarines and missiles." His own job was to track the edges of the radioactive cloud produced during the atmospheric tests by driving along the roads in a car, without the benefit of protective clothing.

"We found it by a meter sticking out the window," he says, his eyes blazing. "And then we found the 'hotline' — the hottest, highest reading of the cloud — and they wanted to know where that was. And then the other edge — turn around and go back through it again. The sunglasses I had were brown sunglasses, and I could see the cloud when other guys couldn't see it."

O'Neill runs his hand through his thinning hair. "I didn't wear a hat in those days, and our hair would be radioactive," he says. "Sometimes you'd have dust on your clothes that they could detect. They'd clean us off, decon us. Yeah, it was beautiful."

Because it is a Friday afternoon, and the hour is late, and there is no one else here, he takes out a clear canister to show me his greatest treasure, a shard of melted brown sludge formed at 5:30 A.M. on June 7, 1945.

"That's white sand that's been turned into green glass by the heat," he explains. I look curiously at the sludge, which is known as trinitite — the material formed in the first nuclear blast at Alamogordo, New Mexico. Inside the glass are tiny air bubbles dating back to the exact moment when the atomic world was born. O'Neill holds the glass up to the light.

"It's reading ten to fifteen MR per hour, as I recall," he says.

I ask him what else he has.

"I stole a nut," he says. "And a camera lens." At that moment O'Neill's wife, Melva, whom he met twenty years ago when the United States conducted a nuclear shot in a salt dome in Carlsbad, New Mexico, enters the room and overhears the last part of our conversation.

"Don't tell him that," she says.

"Well," he says gently, "it's not radioactive anymore."

Notes

1. Honoring mountains, rivers, famous scientists, fish, birds, small mammals, automobiles, Indian tribes, cheeses, wines, cocktails, fabrics, nautical terms, ghost towns, and games of chance and skill, the names of these tests range from Bandicoot, Marshmallow, Harlem, Barracuda, Sardine, Pike, and Minnow to Gumdrop, Gouda, Gruyere, Chevre, Asiago, Reblochon, Pile Driver, Discus Thrower, Double Play, Emerson, Buff, Fawn, Effendi, Akbar, Commodore, Knickerbocker, Vito, Stanley, Lexington, Hupmobile, Noggin, Wineskin, Horehound, Diamond Dust, Seafoam, and Freezeout.

2. At present, the United States has approximately 24,000 intact plutonium pits, each of which could detonate in a blast a little smaller than the one at Hiroshima. Of these, about 10,600 pits are inside active weapons. The rest are kept in sealed bunkers at the Pantex nuclear weapons assembly plant. For most of the Cold War, pits were made at the Rocky Flats plant in Colorado, which was shut down in 1989 after it was raided by the FBI for gross violations of environmental law in the middle of a production run for the 450-kiloton W-88 warhead for the Trident II missile system. Since the raid on Rocky Flats, the United States has lacked the capacity to manufacture "certified" pits for its nuclear weapons. According to the National Nuclear Security Administration's most recent budget submission, the agency will spend $132 million in 2006 to manufacture six pits at a temporary facility at Los Alamos National Laboratory, which is scheduled to deliver, sometime in 2007, at least one pit worthy of being deemed "certified."

3. In 1965 the Atomic Energy Commission encouraged the formation of the CER Geonuclear Corporation, a joint venture of Continental Oil Company, EG&G, and REECo. The cost of a two-megaton nuclear device for private use was set by the AEC at $2 million, a price that included arming and firing the weapon but did not include the cost of preparing the site and digging the hole. In December 1967, in what was called Project Gasbuggy, the Geonuclear Corporation exploded a nuclear device near Farmington, New Mexico. Gasbuggy was followed by Rulison, a shot done

in concert with the Austral Oil Company out of Houston. A man named Les Donovan actually had the job of going from door to door to inform families that three nuclear bombs would be going off in their neighborhood on behalf of the Austral Oil Company in order to stimulate the flow of gas through the surrounding rock.

4. The pit is indeed the approximate size of a softball and is typically made of a hollow, round shell of Pu-239. Alloyed plutonium, like steel, can be cast, pressed, and machined into spheres.

5. Plutonium pits are coated with nickel or gold to contain alpha radiation, which is light enough to be stopped by rubber gloves or a sheet of paper. They are slightly warm to the touch.

6. Seaborg named the new element plutonium after the planet Pluto, which had been discovered just eleven years earlier. Plutonium-239, a variant of the element Seaborg discovered, proved more likely to fission than uranium, making it the perfect fuel for a nuclear bomb. Seaborg chose the letters "Pu" for the new element as a childish joke. All of the plutonium now on earth is man-made.

7. From soup to nuts, the cost of developing a single new warhead might range from two million to five million dollars.

8. Getting American troops used to the idea of marching through radioactive blast zones was an important part of the atmospheric program. At the Desert Rock 1 test, in November 1951, more than six hundred men would be interviewed by psychologists from the Human Resources Research Organization, a new government contractor. Worried that "inclination to panic in the face of AW [atomic warfare] and RW [radiological warfare] may prove high," the Joint Panel on the Medical Aspects of Atomic Warfare proposed to "increase research efforts in the scientific study of panic and its results" by exposing "as many men as possible" to nuclear tests. During Operation Jangle, teams of soldiers were ordered to walk over ground zero for one hour to test the effectiveness of protective clothing, while others rode tanks through contaminated areas to check the shielding effects of the armor.

JOSH SCHOLLMEYER

Lights, Camera, Armageddon

FROM *Bulletin of the Atomic Scientists*

WHEN THE FILMMAKERS OF *Thirteen Days* told Graham Allison that they intended to transform President John F. Kennedy's appointments secretary, Kenneth O'Donnell, into a major player during the Cuban missile crisis, he objected. Immediately.

"That's stupid," Allison, author of *Essence of Decision: Explaining the Cuban Missile Crisis,* remembers telling them. "I don't relate to it."

"But how many people read your book?" they responded.

So begins the quandary: experts get it right; Hollywood delivers the crowds. A happy marriage of the two qualifies as an exception. For every *Thirteen Days* — a film hailed by historians and critics alike — there are countless potboiler thrillers that twist and distort reality, all in the name of popcorn purchases and ticket sales.

Experts like Allison can see past the showbiz gloss — the omissions, distortions, and casual disregard for even the most basic laws of physics. On planet Hollywood, nuclear power plants melt down with the push of a button, and nuclear warheads can be disarmed with the snip of a wire. But what goes through the minds of moviegoers when Ben Affleck struts unharmed along the fringes of a nuclear explosion in *The Sum of All Fears*? Such scenarios linger in the public's collective consciousness. Popular culture resonates. Allison recognizes it at Harvard University, where he teaches.

"Why did the CIA murder Kennedy?" students often ask him.

"Where did you get this idea?" a befuddled Allison responds.

"I saw it in *JFK,*" they inform him.

Superficially, many policy wonks and activists dismiss films like

The Sum of All Fears as vapid Hollywood entertainment, claiming that their influence lands at the margins. But they intuitively understand popular culture's pull: that the size of *The Peacemaker*'s audience is of a significantly higher order of magnitude than C-Span's. They see in the entertainment industry an opportunity to inform the public and to advance their agendas. Yet as they inevitably learn, Hollywood has an agenda all its own.

A Pox upon TV Viewers

The British director Daniel Percival wanted *Smallpox* to be educational. He researched the virus extensively before writing and filming the BBC Two faux documentary about a major bioterror attack on a Western city. (The film was released in Britain in 2002, but because of post–9/11 sensitivities it did not air in the United States until January 2, 2005.)

His mission was twofold: he wanted to alert the public to the consequences of a bioterror event and to create an engaging film that would be "an emotional as well as intellectual journey."

Unfortunately, *Smallpox* was neither. The film played more like a training video for the World Health Organization (WHO) and for emergency responders — Percival alleges that the film is used for this exact purpose — than a riveting docudrama. And despite Percival's prolific research, *Smallpox* contained inaccuracies. In the film a lone terrorist of unknown origin and affiliation infects himself with smallpox and then saunters around the city streets spreading the virus.

"I got a copy of it not long before it was going to air [in Britain], and I exploded," says D. A. Henderson, who led the WHO's global smallpox eradication campaign. "I said, 'No, this is wrong!'"

During preproduction Henderson had spoken with one of the film's producers. After viewing the finished product, he called the producer, detailing his concerns. "We'll see what we can do," the producer told him. The next version of the film Henderson saw contained some changes, but errors remained. Most notably, the terrorist appears to be in peak physical health as he trolls the streets, supposedly infecting 10 out of the 150 people he comes in contact with.

Henderson was most angered by a scene contending that a light tap on the head would be enough to transfer the virus. "It's just not all that good of a spreader," he says. "They're infected, but they're not contagious until they actually develop symptoms — headaches, abdominal pain, high fever. The infectiousness you have is directly correlated to how sick you are. The really sick ones don't move."

The Centers for Disease Control and Prevention sent out physicians to debunk the film's premise, but as it turns out, they and Henderson needn't have worried. *Smallpox* failed to register among U.S. audiences, in part because it couldn't compete on TV with college football. Plus *Smallpox* (which offered the ominous, ratings-driven tag line "It's All True. It Just Hasn't Happened Yet") had already become dated. The emergency response to bioterrorism in the United States substantially improved after 9/11 and the anthrax attacks. *Smallpox* vaccines were now readily accessible, with the United States stockpiling enough for everyone in the country.

Despite the criticism, Percival remained undeterred. Greatly influenced by Peter Watkins's 1965 post–nuclear war opus *The War Game* — a graphic docudrama that revealed the inadequacies of Britain's civil defense programs — he was determined to make an evocative film with a strong public service message behind it.

Percival turned his attention to what he saw as an equally pressing threat: dirty bombs. This time, in making his film *Dirty War*, he vetted the research more closely. He attacked it like a journalist, verifying every line of dialogue with BBC News. Unofficially, he consulted all the relevant British government agencies. For every fact he required three independent sources. "If you're going to tackle these issues and try to have a genuine impact, you've got to be unimpeachable," Percival says. "Otherwise the film fails to serve a purpose."

His goal was nothing less than to change public policy. "It became clear in my mind after making *Smallpox* and before I made *Dirty War* that we concentrated all our energies, resources, and planning on delivering a counterpunch for 9/11 rather than thinking about how the hell we are going to absorb the punch when it comes back," he says.

Dirty War definitely unsettles. Islamic extremists smuggle radioactive material into Britain and begin assembling explosive delivery devices, which they intend to detonate in London. By happenstance, London police and British intelligence discover the cells planning the attack, but not before the terrorists detonate one of the dirty bombs.

Chaos and panic prevail. Emergency responders are overwhelmed. Firemen attempt to douse the fires at ground zero despite cumbersome protective suits and high levels of radiation. Nearby, 300,000 Londoners are cordoned off in decontamination zones. Police attempt to secure these zones, imploring those inside not to leave until they've been sufficiently decontaminated. Not many stay.

As Percival hoped, *Dirty War* got people talking, first in Britain, where it aired on the BBC, and then in the United States, where it aired originally on HBO and later on PBS. A panel discussion convened by PBS featured an impressive dais: New York City's police commissioner, Ray Kelly; former New Hampshire Republican senator Warren Rudman; former deputy Homeland Security adviser Richard Falkenrath; and the Nuclear Threat Initiative's vice president for biological programs, Margaret Hamburg.

On the program Kelly admitted that he screened *Dirty War* for his department's top brass: "It was sobering. It was realistic. It was well done, and obviously food for thought."

Unlike *Smallpox, Dirty War* impressed those in the field. "I came down with the same opinion as many other people," says Joel Lubenau, an expert on radiological dispersion devices at the Center for Nonproliferation Studies at the Monterey Institute of International Studies. "In terms of technical aspects, I saw very little to criticize."

If anything, *Dirty War* was faulted with understating the threat. Percival wove an intricate terrorist conspiracy to tell his tale. In truth, a much less elaborate scenario with fewer terrorists could unfurl an equally devastating attack. "The terrorists in the movie used a more difficult means of bringing the bomb to bear on the target," Falkenrath remarked during the panel discussion. "They brought in the material from [abroad]. There's no reason to do that. There's ample quantities of radioactive material in the United Kingdom and in the United States that they could acquire here and use."

Dirty War became Percival's *War Game.* "There's no reason why you can't handle important and strong messages in a dramatic form," the filmmaker explains. "We're not used to getting information in that way, but judging by the response to *Dirty War* it's clear that it worked. It opens a new way of using the dramatic narrative to deal with real issues."

Wish upon a Movie Star

Jessica Stern viewed *The Peacemaker*— a 1997 movie about nuclear terrorism *very* loosely based on her involvement with securing Russian nuclear weapons and material after the fall of the Soviet Union — as a sort of cinematic op-ed piece. "I thought it was a really useful service," Stern says, "which was getting people to pay attention to a problem that was, at that point, quite underfunded."

Stern unwittingly became the basis for Nicole Kidman's character in *The Peacemaker* during an interview with Leslie Cockburn, an investigative reporter intending to write an article for *Vanity Fair* about the Clinton administration's inaction regarding loose nukes and nuclear smuggling.

Some time later Stern received a call from the film studio DreamWorks, asking her to meet with a screenwriter. When she did, she was startled to learn that Cockburn's article had morphed into a film treatment and featured a lead character named Jessica Stern. The screenwriter asked Stern to sign a release. She refused, concerned that it might harm her career.

Ultimately, DreamWorks changed the character's name to Dr. Julia Kelly and altered her background enough that Stern felt comfortable serving as a consultant. On the set the producers asked her to teach Kidman how to disable a nuclear weapon for the film's penultimate scene. This Hollywood fantasy Stern would not indulge. "I have no idea how to disable a nuclear device, and I'm not going to pretend I know," she told them, leaving the fate of the world to Kidman.

A lot was riding on *The Peacemaker.* As the first film from DreamWorks — an upstart studio formed by the entertainment industry heavyweights Steven Spielberg, Jeffrey Katzenberg, and David Geffen — a gigantic box office gross was expected. Because of this, accuracy became a tertiary impulse.

Stern understood: "I don't expect it to be a textbook. I'm sorry for anybody who thinks they're going to teach themselves how to disable or detect nuclear devices by watching a Hollywood film. At the same time, the overall message is very important for people to hear."

To provide context for this message, Stern attended the film's press junket and weathered questions such as "Is George Clooney cute?" She took her message to the tabloid celebrity news program *Entertainment Tonight* — a rare, if not unique, opportunity for a National Security Council staffer to directly engage a mainstream audience.

While Stern stumped for the issues, Kidman and Clooney — *The Peacemaker*'s other star — stumped for the film. Scrounging for every budget dollar, Laura Holgate, the former manager of the Defense Department's United States–Russia Cooperative Threat Reduction program — whom DreamWorks also consulted during preproduction of *The Peacemaker* — hoped that Kidman or Clooney would grow concerned about loose nukes during their research for the film and latch on to the issue as their pet political cause. She imagined that with someone famous campaigning for the issue, a privatized effort might develop.

Kidman and Clooney passed. Loose nukes were a cause before their time, and no celebrity came forward until 2000, when CNN's founder, the rabble-rousing philanthropist Ted Turner, created the Nuclear Threat Initiative — Holgate's current employer — to address worldwide concerns about the security of weapons of mass destruction. "I won't call it an accountability gap," Holgate says, "but there's a missed opportunity here for Hollywood as a structure and for stars as individuals to learn for themselves from what they've been doing and make that a basis of some advocacy."

Waste Not, Want Not

Martin Sheen almost qualifies more as an activist than as an actor. So the nuclear industry was understandably anxious when it discovered that the NBC drama *The West Wing,* which stars Sheen as a powerful Democratic president, was planning an April 2002 plot line detailing an accident in Idaho involving a truck carrying spent uranium fuel rods.

The West Wing carries a gravitas not many network dramas possess. Then at the critical and commercial peak of its run, the series reached millions of homes each week. This particular episode especially blurred personal politics with art. The former Clinton White House press secretary Dee Dee Myers helped craft the nuclear-waste story arc, and Sheen has long been a vocal opponent of Yucca Mountain, the proposed nuclear-waste depository in Nevada, and has protested at the site many times.

With much political and PR capital at stake, the spinmeisters pounced immediately. Those in Nevada opposed to Yucca Mountain saw the episode as free national advertising. "The episode was extremely helpful in raising awareness about waste-shipping issues," Nevada's Republican senator John Ensign told the *Bulletin*. "The free media coverage was a great addition to paid advertising and showed that Yucca Mountain has been and will always be a dangerous policy for the United States."

Angry that biased voices were framing the waste debate, the nuclear industry attacked the episode's accuracy. The day after it aired, Jack Edlow and David Blee, executives from two nuclear transportation firms, staged a news conference in Washington, D.C., to refute the episode point by point. During the news conference they played a tape of the show's nuclear-waste segment, analyzing it for what they maintained were gross inaccuracies. "This [episode] might as well have been produced in Las Vegas," Blee said. "It is part of a calculated campaign being waged by opponents of Yucca Mountain."

"We want policymakers and the public to have an opportunity to hear from people who really know what they're talking about, rather than guys like Martin Sheen who pretend to be things," Edlow added.

In Nevada the industry's reaction fed into the anti–Yucca Mountain movement's spin. Without the industry's ire, the episode might have had less impact nationally — the nuclear waste story line was buried beneath a soapy alcoholism subplot and a parable about taxes. But those in the industry sparked controversy and, in doing so, intensified media coverage. "You can do the least little thing, and they just knee-jerk to it," says Bob Loux, the director of the Nevada Agency for Nuclear Projects. "They overreact to the point that it really helps you out."

In the decades following the nuclear accidents at Three Mile Island and Chernobyl, the nuclear industry has tried to become more savvy about promoting itself. The executives understand that the public is leery about all things nuclear and have tried to assail these fears with multimillion-dollar advertising campaigns, lobbying efforts, and attempts at courting the media. They devised "truth squads" of scientists and engineers to present the industry's position on nuclear matters and a "nuclear energy branding campaign" to better express their viewpoint.

In Washington and in the news media, these strategies have helped make some inroads. But those in the entertainment industry remain unconvinced. What the nuclear industry fails to grasp is that television programs and movies deal in worst-case scenarios. To a screenwriter, fiery nuclear waste and potentially catastrophic explosions provide an exciting dramatic structure — unlike the safe passage of nuclear waste from a U.S. nuclear plant to Yucca Mountain.

And if those catastrophic scenarios push the envelope of believability, so much the better. A story arc in the latest season of the Fox Network's hit series *24* features our hero, government agent Jack Bauer, in a race against time to thwart Islamic terrorists who have stolen a remote "override device" capable of triggering meltdowns at all of the nation's nuclear power plants.

The Nuclear Energy Institute (NEI), the nuclear industry's lobbying arm, launched a PR counterattack. "I've come to the conclusion that if nuclear plants didn't exist, Hollywood would have to invent them," Steve Kerekes, a spokesman for NEI, publicly complained. Skip Bowman, the president and CEO of NEI, urged Fox Entertainment Group to issue a disclaimer, since the plot line was "unrealistic and holds the potential to needlessly frighten viewers who are unaware that no such remote override device exists." (Fox had previously televised a disclaimer on the program, reminding viewers that the "American Muslim community stands firmly beside their fellow Americans in denouncing and resisting all forms of terrorism.") The NEI's Web log reports that while Fox appreciated the nuclear industry's concerns, it would not issue a disclaimer because *24* is a purely fictional program.

"The very fact that the nuclear industry finds it very difficult to promote and find positive images of itself speaks to an innate suspicion, if not hostility, toward the genie being out of the bottle," says Mick Broderick, the author of *Nuclear Movies* and the forthcoming

Entertaining Armageddon: Representing the Unthinkable. "And we don't like the look of it."

Box Office Bomb

Cold War films such as *Fail-Safe, On the Beach,* and *Dr. Strangelove* are cultural time capsules — remnants of a duck-and-cover era when the threat of nuclear apocalypse permeated public discourse. Even the radiation-spawned mutants that populated the B movies of the period spoke of the public's anxiety over the destructive potential unleashed by the atomic age.

Today we've entered the era of the friendly, functional cinematic nukes. These nuclear weapons aren't the proverbial "destroyer of worlds," but saviors of humanity. They possess utility and a higher purpose. Twice they've thwarted a giant asteroid from slamming into the earth (*Deep Impact* and *Armageddon*), and once they helped reset the rotation of its core (*The Core*).

Amazingly, these films reflect the zeitgeist. Polls indicate that nuclear weapons are no longer the number one public fear. (U.S. mayors listed traditional crime and fires as a greater concern to their cities and constituents than a nuclear threat in a survey eight months after 9/11.) "Nuclear weapons have been recuperated during the 1990s," Broderick says. "These weapons were redundant in the post–Cold War period, but now they have capital because they help us liberate society from asteroids, comets, alien invasion, what have you. Previously they were only there to destroy society."

Indeed, the public now cites biological terrorism as its number one fear. Jessica Stern, writing in the Winter 2002/2003 issue of the journal *International Security,* wryly noted that although malaria kills one million people per year and Ebola and Marburg have collectively killed fewer than a thousand, "it is Ebola and Marburg that have inspired terrifying books and movies. We respond to the likelihood of death in the event the disease is contracted, rather than the compound probability of contracting the disease and succumbing to its effects." It is the exotic nature of such diseases, she argues, particularly in industrialized societies, that increases "their hold on our imagination" and increases the "dread factor."

Bioterrorism engenders fear also because it is ubiquitous (a contagion knows no geographic borders) and arbitrary (an anthrax-filled letter can arrive in any mailbox). By contrast, to the extent that nukes remain a source of concern, they have perversely morphed into a "localized threat" — something that can lay waste not to a continent but to a single city unlucky enough to be targeted by terrorists. In *The Sum of All Fears,* a nuclear detonation is reduced to a plot point. Little attention is paid to the thousands of people who would have been incinerated. While Cold War–era films like *Testament* and *The Day After* depicted, in sometimes agonizing detail, the slow deaths resulting from radioactive fallout, the *Sum of All Fears* superagent Jack Ryan stands on the outskirts of ground zero shouting into his (miraculously working) cell phone.

"I remember walking out of there saying, 'Oh jeez,'" recalls Holgate about seeing the film. However, her professional distaste notwithstanding, she is skeptical that such depictions are taken seriously. "James Bond is not where you go for the user's manual for a car," Holgate notes. "These films are part of a genre that people understand is a thriller. It's a story."

Jim Walsh, the executive director of Harvard University's Managing the Atom Project, agrees. "At the end of the day, movies and TV programs really aren't going to be the defining elements of the public's perceptions," he says. "The public perception of radiation and radiation-related events has been decades in the making."

Broderick is not so sure: "These issues do have a resonance that operates not necessarily overtly or consciously."

If it's difficult to gauge the impact that films are having on the public psyche, it may be partly because the very concept of plausibility is now stretched to its limits. In the aftermath of 9/11, the satirical newspaper *The Onion* ran the darkly comical headline AMERICAN LIFE TURNS INTO BAD JERRY BRUCKHEIMER MOVIE. It is an apt description. We now live in an era in which we are loath to dismiss any potential terrorist plot, no matter how elaborate, as impossible. The media bombard us with speculative doomsday scenarios that lack analysis and context. At the policy-making level, the public is confronted with a color-coded terrorist alert system that nobody entirely understands and a government that includes water parks and a miniature golf course among its list of potential terrorist targets.

Against such a backdrop, credible films such as *Dirty War* may struggle to stand out, let alone galvanize the public. Plausible terrorist scenarios that merit increased scrutiny — like a commercial airliner crashing into a power plant — are eclipsed by fantastical plot lines, as in *24*. And those who see Hollywood as a medium for political change will learn that popular culture can only do so much — especially when government still hasn't quite worked out the story line.

MONCEF ZOUALI

Taming Lupus

FROM *Scientific American*

A TWENTY-FOUR-YEAR-OLD WOMAN undergoes medical evaluation for kidney failure and epilepsy-like convulsions that fail to respond to antiepileptic drugs. Her most visible sign of illness, though, is a red rash extending over the bridge of her nose and onto her checks, in a shape resembling a butterfly.

A sixty-three-year-old woman insists on being hospitalized to determine why she is fatigued, her joints hurt, and breathing sometimes causes sharp pain. Ever since her teen years she has avoided the sun, which raises painful blistering rashes wherever her skin is unprotected.

A twenty-year-old woman is surprised to learn from a routine health exam that her urine has an abnormally high protein level — a sign of disturbed kidney function. A renal biopsy reveals inflammation.

Although the symptoms vary, the underlying disease in all three patients is the same — systemic lupus erythematosus, which afflicts an estimated 1.4 million Americans, including one out of every 250 African American women aged eighteen to sixty-five. It may disrupt almost any part of the body: skin, joints, kidneys, heart, lungs, blood vessels, or brain. At times it becomes life-threatening.

Scientists have long known that fundamentally lupus arises from an immunological malfunction involving antibody molecules. The healthy body produces antibodies in response to invaders, such as bacteria. These antibodies latch on to specific molecules on an invader that are sensed as foreign (antigens) and then damage the interloper directly or mark it for destruction by other parts of the immune system. In patients with lupus, however, the body produces antibodies that perceive its own molecules as foreign and then launch an attack targeted to those "self-antigens" on the body's own tissues.

Self-attack — otherwise known as autoimmunity — is thought to underpin many diseases, including type 1 diabetes, rheumatoid arthritis, multiple sclerosis, and possibly psoriasis. Lupus, however, is at an extreme. The immune system reacts powerfully to a surprising variety of the patient's molecules, ranging from targets exposed at the surface of cells to some inside cells and even to some within a further sequestering chamber, the cell nucleus. In fact, lupus is notorious for the presence of antibodies that take aim at the patient's DNA. In the test tube, these anti-DNA "autoantibodies" can directly digest genetic material.

Until recently, researchers had little understanding of the causes of this multipronged assault. But clues from varied lines of research are beginning to clarify the underlying molecular events. The work is also probing the most basic, yet still enigmatic, facets of immune system function: the distinction of self from nonself; the maintenance of self-tolerance (nonaggression against native tissues); and the control over the intensity of every immune response. The discoveries suggest tantalizing new means of treating or even preventing not only lupus but also other autoimmune illnesses.

Some Givens

One thing about lupus has long been clear: the autoantibodies that are its hallmark contribute to tissue damage in more than one way. In the blood, an autoantibody that recognizes a particular self-antigen can bind to that antigen, forming a so-called immune complex, which can then deposit itself in any of various tissues. Autoantibodies can also recognize self-antigens already in tissues and generate immune complexes on-site. Regardless of how the complexes accumulate, they spell trouble.

For one, they tend to recruit immune system entities known as complement molecules, which can directly harm tissue. The complexes, either by themselves or with the help of the complement molecules, also elicit an inflammatory response. This response involves an invasion by white blood cells, which attempt to wall off and destroy any disease-causing agents. Inflammation is a protective mechanism, but if it arises in the absence of a true danger or goes on for too long, the inflammatory cells and their secretions can harm the tissues they are meant to protect. Inflammation can additionally involve the abnormal proliferation of cells native to an affected tissue, and this cellular excess can disrupt the normal functioning of the tissue. In the kidney, for instance, immune complexes can accumulate in glomeruli, the organ's blood-filtering knots of capillary loops. Excessive deposition then initiates glomerulonephritis, a local inflammatory reaction that can lead to kidney damage.

Beyond inciting inflammation, certain lupus autoantibodies do harm directly. In laboratory experiments, they have been shown to bind to and then penetrate cells. There they become potent inhibitors of cellular function.

The real mystery about lupus is what precedes such events. Genetic predisposition seems to be part of the answer, at least in some people. About 10 percent of patients have close blood relatives with the disease, a pattern that usually implies a genetic contribution. Moreover, investigators have found greater lupus concordance — either shared lupus or a shared absence of it — in sets of identical twins (who are genetically indistinguishable) than in sets of fraternal twins (whose genes generally are no more alike than those of other pairs of siblings).

Genetic Hints

Spurred by such findings, geneticists are hunting for the genes at fault, including those that confer enhanced susceptibility to the vast majority of patients who have no obvious family history of the disease. Knowledge of the genes, the proteins they encode, and the normal roles of these proteins should one day help clarify how lupus develops and could point to ways to better control it.

In mice prone to lupus, the work has identified more than thirty fairly broad chromosomal regions associated to some extent either with lupus or with resistance to it. Some are tied to specific elements of the disease. One region, for example, apparently harbors genes that participate in producing autoantibodies that recognize components of the cell nucleus (although the region itself does not encode antibodies); another influences the severity of the kidney inflammation triggered by lupus-related immune complexes.

In human lupus, the genetic story may be even more mind-boggling. An informative approach scans DNA from families with multiple lupus patients to identify genetic features shared by the patients but not by the other family members. Such work has revealed a connection between lupus and forty-eight chromosomal regions. Six of those regions (on five different chromosomes) appear to influence susceptibility most. Now investigators have to identify the lupus-related genes within those locales.

Already it seems fair to conclude that multiple human genes can confer lupus susceptibility, although each gene makes only a hard-to-detect contribution on its own. And different combinations of genes might lay the groundwork for lupus in different people. But clearly, single genes are rarely, if ever, the primary driver; if they were, many more children born to a parent with lupus would be stricken. Lupus arises in just about 5 percent of such children, and it seldom strikes in multiple generations of a family.

Many Triggers

If genes alone rarely account for the disease, environmental contributors must play a role. Notorious among these is ultraviolet light. Some 40 to 60 percent of patients are photosensitive: exposure to sunlight, say for ten minutes at midday in the summer, may suddenly cause a rash. Prolonged exposure may also cause flares, or increased symptoms. Precisely how it does so is still unclear. In one scenario, ultraviolet irradiation induces changes in the DNA of skin cells, rendering the DNA molecules alien (from the viewpoint of the body's immune defenses) and thus potentially antigenic. At the same time, the irradiation makes the cells prone to breakage, at which point they will release the antigens, which can then provoke an autoimmune response.

Environmental triggers of lupus also include certain medications, among them hydralazine (for controlling blood pressure) and procainamide (for irregular heartbeat). But symptoms usually fade when the drugs are discontinued. In other cases an infection, mild or serious, may act as a lupus trigger or aggravator. One suspect is the Epstein-Barr virus, perhaps best known for causing infectious mononucleosis, or "kissing disease." Even certain vaccines may provoke a lupus flare. Yet despite decades of research, no firm proof of a bacterium, virus, or parasite that transmits lupus has been put forth. Other possible factors include diets high in saturated fat, pollutants, cigarette smoking, and perhaps extreme physical or psychological stress.

Perils of Cell Suicide

Another line of research has revealed cellular and molecular abnormalities that could well elicit or sustain autoimmune activity. Whether these abnormalities are usually caused more by genetic inheritance or by environmental factors remains unknown. People may be affected by various combinations of influences.

One impressive abnormality involves a process known as apoptosis, or cell suicide. For the body to function properly, it has to continually eliminate cells that have reached the end of their useful life or turned dangerous. It achieves this pruning by inducing the cells to make proteins that essentially destroy the cell from within — such as by hacking to pieces cellular proteins and the chromosomes in the nucleus. But the rate of apoptosis in certain cells — notably, the B and T lymphocytes of the immune system — is excessive in those who have lupus.

When cells die by apoptosis, the body usually disposes of the remains efficiently. But in those with lupus, the disposal system seems to be defective. This double whammy of increased apoptosis and decreased clearance can promote autoimmunity in a fairly straightforward way: if the material inside the apoptotic cells is abnormal, its ejection from the cells in quantity could well elicit the production of antibodies that mistakenly perceive the aberrant material as a sign of invasion by a disease-causing agent. And such antibody production is especially likely if the ejected material, rather than being removed, accumulates enough to call attention to itself.

Unfortunately, the material that spills from apoptotic cells of those with lupus, especially the chromosomal fragments, is often abnormal. In healthy cells, certain short sequences of DNA carry methyl groups that serve as tags controlling gene activity. The DNA in circulating immune complexes from lupus patients is undermethylated. Scientists have several reasons to suspect that this methylation pattern might contribute to autoimmunity. In the test tube, abnormally methylated DNA can stimulate a number of cell types involved in immunity, including B lymphocytes, which, when mature, become antibody-spewing factories. (Perhaps the body misinterprets these improperly methylated stretches as a sign that a disease-causing agent is present and must be eliminated.) Also, certain drugs known to cause lupus symptoms lead to undermethylation of DNA in T cells, which leads to T-cell autoreactivity in mice.

All in all, the findings suggest that apoptotic cells are a potential reservoir of autoantigens that are quite capable of provoking an autoantibody response. In further support of this idea, intravenous administration of large quantities of irradiated apoptotic cells is able to induce autoantibody synthesis in normal mice.

Hence, part of the underlying process leading to the formation of destructive immune complexes may involve the body's production of foreign-seeming antigens, which cause the body to behave as if tissues bearing those antigens were alien and threatening. But other work indicates that in addition the B lymphocytes of lupus patients are inherently deranged; they are predisposed to generate autoantibodies even when the self-molecules they encounter are perfectly normal. In other words, the mechanisms that should ensure self-tolerance go awry.

Deranged Cells

The problem seems to stem mostly from signaling imbalances within B cells. In the healthy body, a B cell matures into an antibody-secreting machine — known as a plasma cell — only after antibodylike projections on the B cell's surface (B-cell receptors) bind to a foreign antigen. If a B cell instead attaches to a self-component, this binding normally induces the cell to kill itself, to retreat into a nonresponsive (anergic) state or to "edit" its receptors until they can no longer recognize the self-antigen.

Whether the cell responds appropriately depends in large measure on the proper activity of the internal signaling pathways that react to external inputs. Mouse studies show that even subtle signaling imbalances can predispose animals to produce antibodies against the self. And various lines of evidence indicate that certain signaling molecules (going by such names as Lyn, CD45, and SHP-1) on and in B cells of patients with lupus are present in abnormal amounts.

Other work suggests that it is not only the B cells that are deranged. For a B cell to become an antibody maker, it must do more than bind to an antigen. It must also receive certain stimulatory signals from immune system cells known as helper T lymphocytes. Helper cells of lupus patients are afflicted by signaling abnormalities reminiscent of those in the B cells. The T cell aberrations, though, may lead to autoantibody production indirectly — by causing the T cells to inappropriately stimulate self-reactive B cells.

All theorizing about the causes of lupus must account not only for the vast assortment of autoantibodies produced by patients but also for another striking aspect of the disease: the disorder is ten times more common in women than in men. It also tends to develop earlier in women (during childbearing years). This female susceptibility — a pattern also seen in some other autoimmune diseases — may stem in part from greater immune reactivity in women. They tend to produce more antibodies and lymphocytes than males and, probably as a result, to be more resistant to infections. Among mice, moreover, females reject foreign grafts more rapidly than the males do. Perhaps not surprisingly, sex hormones seem to play a role in this increased reactivity, which could explain why, in laboratory animals, estrogens exacerbate lupus and androgens ameliorate it.

Estrogens could pump up immune reactivity in a few ways. They augment the secretion of prolactin and growth hormone, substances that contribute to the proliferation of lymphocytes, which bear receptor molecules sensitive to estrogens. Acting through such receptors, estrogens may modulate the body's immune responses and may even regulate lymphocyte development, perhaps in ways that impair tolerance of the self.

Toward New Therapies

Those of us who study the causes of lupus are still pondering how the genetic, environmental, and immunological features that have been uncovered so far collaborate to cause the disease. Which events come first, which are most important, and how much do the underlying processes differ from one person to another? Nevertheless, the available clues suggest at least a partial scenario for how the disease could typically develop.

The basic idea is that genetic susceptibilities and environmental influences may share responsibility for an impairment of immune system function — more specifically, an impairment of the signaling within lymphocytes and possibly within other cells of the immune system, such as those charged with removing dead cells and debris. Faulty signaling, in turn, results in impaired self-tolerance, accelerated lymphocyte death, and defective disposal of apoptotic cells and the self-antigens they release. Abundantly available to the body's unbalanced immune surveillance, the antigens then misdirect the immune system, inducing it to attack the self.

Drugs do exist for lupus, but so far they focus on dampening the overall immune system. In other words, they are nonspecific: instead of targeting immunological events underlying lupus in particular, they dull the body's broad defenses against infectious diseases. Corticosteroids, for instance, reduce inflammation at the cost of heightening susceptibility to infections.

The challenge is to design new drugs that prevent autoimmune self-attacks without seriously hobbling the body's ability to defend itself against infection. To grasp the logic of the approaches being attempted, it helps to know a bit more about how helper T cells usually abet the transformation of B cells into vigorous antibody makers.

First, the helper cells themselves must be activated, which occurs through interactions with so-called professional antigen-presenting cells (such as macrophages and dendritic cells). These antigen presenters ingest bacteria, dead cells, and cellular debris; chop them up; join the fragments to larger molecules (called MHC class II molecules); and display the resulting MHC-antigen complexes on the cell surface. If the receptor on a helper T cell recognizes a complex and binds to it, the binding conveys an antigen-specific signal into the T cell. If, at the same time, a certain T-cell projection near the receptor links to a particular partner (known as a B7 molecule) on the antigen-presenting cell, this binding will convey an antigen-independent, or costimulatory, signal into the T cell. Having received both messages, the T cell will switch on; that is, it will produce or display molecules needed to activate B cells and will seek out those cells.

Like the professional antigen-presenting cells, B cells display fragments of ingested material — notably fragments of an antigen they have snared — on MHC class II molecules. If an activated helper T cell binds through its receptor to such a complex on a B cell, and if the T and B cells additionally signal each other through costimulatory surface molecules, the B cell will display receptors for small proteins called cytokines. These cytokines, which are secreted by activated helper T cells, induce the B cell to proliferate and mature into a plasma cell, which dispatches antibodies that specifically target the same antigen recognized by the coupled B and T cells.

Of course, any well-bred immune response shuts itself off when the danger has passed. Hence, after an antigen-presenting cell activates a helper T cell, the T cell also begins to display a "shutoff" switch known as CTLA-4. This molecule binds to B7 molecules on antigen-presenting cells so avidly that it links to most or even all of them, thereby putting a brake on any evolving helper T- and, consequently, B-cell responses.

One experimental approach to treating lupus essentially mimics this shutoff step, dispatching CTLA-4 to cap over B7 molecules. In mice prone to lupus, this method prevents kidney disease from progressing and prolongs life. This substance is beginning to be tested in lupus patients; in those with psoriasis, initial clinical trials have shown that the treatment is safe.

A second approach would directly impede the signaling between helper T cells and B cells. The T-cell molecule that has to "clasp hands" with a B-cell molecule to send the needed costimulatory signal into B cells is called CD154. The helper cells of lupus patients show increased production of CD154, and in mice prone to the disease, antibodies engineered to bind to CD154 can block B-cell activation, preserve kidney function, and prolong life. So far, early human tests of different versions of anti-CD154 antibodies have produced a mixture of good news and bad. One version significantly reduced autoantibodies in the blood, protein in the urine, and certain symptoms, but it also elicited an unacceptable degree of blood-clot formation. A different version did not increase thrombosis but worked poorly. Hence, no one yet knows whether this approach to therapy will pan out.

A third strategy would interfere with B-cell activity in a different way. Certain factors secreted by immune system cells, such as the cytokine BAFF, promote cell survival after they bind to B cells. These molecules have been implicated in various autoimmune diseases, including lupus and its flares; mice genetically engineered to overproduce BAFF or one of its three receptors on B cells develop signs of autoimmune disease, and BAFF appears to be overabundant both in lupus-prone mice and in human patients. In theory, then, preventing BAFF from binding to its receptors should minimize antibody synthesis. Studies of animals and humans support this notion. In mice a circulating decoy receptor, designed to mop up BAFF before it can find its true receptors, alleviates lupus and lengthens survival. Findings for a second decoy receptor are also encouraging. Human trials are in progress.

Targeting other cytokines might help as well. Elevated levels of interleukin-10 and depressed amounts of transforming growth factor (TGF) beta are among the most prominent cytokine abnormalities reported in lupus, and lupus-prone mice appear to benefit from treatments that block interleukin-10 or boost TGF beta. Taking a different tack, investigators studying various autoimmune conditions are working on therapies aimed specifically at reducing the numbers of B cells. An agent called rituximab, which removes B cells from circulation before they are able to secrete antibodies, has shown promise in early trials in patients with systemic lupus.

Some other therapies under investigation include molecules designed to block production of anti-DNA autoantibodies or to induce those antibodies to bind to decoy compounds that would trap them and provoke their degradation. An example of such a decoy is a complex consisting of four short DNA strands coupled to an inert backbone. Although the last idea is intriguing, I have to admit that the effects of introducing such decoys are apt to be complex.

Certain cytokines might be useful as therapies, but these and other protein drugs could be hampered by the body's readiness to degrade circulating proteins. To circumvent such problems, researchers are considering gene therapies, which would give cells the ability to produce useful proteins themselves. DNA encoding transforming growth factor beta has already been shown to treat lupus in mice, but too few tests have been done yet in humans to predict how useful the technique will be. Also, scientists are still struggling to perfect gene therapy techniques in general.

As treatment-oriented investigators pursue new leads for helping patients, others continue to probe the central enigmas of lupus. What causes the aberrant signaling in immune cells? And precisely how does such deranged signaling lead to autoimmunity? The answers may well be critical to finally disarming the body's mistaken attacks on itself.

Contributors' Notes

Other Notable Science and Nature Writing of 2005

Contributors' Notes

Natalie Angier is a Pulitzer Prize–winning journalist and a *New York Times* contributing writer. She is the author of the bestseller *Woman: An Intimate Geography,* which has been translated into nineteen languages; *The Beauty of the Beastly;* and *Natural Obsessions.* Angier was the editor of *The Best American Science and Nature Writing 2002* and has written for *Time, Discover,* the *Atlantic Monthly, Natural History, Preservation,* the *American Scholar, Orion,* and many other magazines. In 2003 she won the Freedom from Religion Foundation's Emperor Has No Clothes Award for her *New York Times Magazine* article "Confessions of a Lonely Atheist." Angier lives in Takoma Park, Maryland, with her husband, Rick Weiss, a science reporter for the *Washington Post,* and their daughter, Katherine.

Drake Bennett is the staff writer for the *Boston Globe* Ideas section. His work has also appeared in the *New York Times, Wired,* the *American Prospect,* the *Boston Review,* and other periodicals. He lives in Cambridge, Massachusetts, with his wife, Rebecca.

Larry Cahill received his Ph.D. in neuroscience in 1990 from the University of California, Irvine, where he worked in the laboratory of James L. McGaugh. After two years of postdoctoral work in Germany using imaging techniques to study learning and memory in rodents, he returned to Irvine to extend his studies of emotional memory to the human animal, on which his research now focuses. He is an associate professor in the Department of Neurobiology and Behavior and a fellow of the Center for the Neurobiology of Learning and Memory at UC Irvine.

Michael Chorost was born deaf in 1964 during an epidemic of rubella and did not learn to talk until he got hearing aids at the age of three and a half.

He received a B.A. in English from Brown in 1987 and a Ph.D. in educational technology from the University of Texas, Austin, in 2000. Between 1999 and 2004 he worked as a technical writer, and he now teaches at the University of San Francisco. *Rebuilt: How Becoming Part Computer Made Me More Human* (2005), his memoir of going completely deaf and getting his hearing back with a cochlear implant, ranges through engineering, computer science, science fiction, and literature to explore the idea of the cyborg: the creative fusion of human and computer. Chorost's Web site is http://www.michaelchorost.com.

Daniel C. Dennett is University Professor at Tufts University in Medford, Massachusetts, and the codirector of the Center for Cognitive Studies there. He is the author of *Breaking the Spell* (2006).

Frans B. M. de Waal is the C. H. Candler Professor of Primate Behavior at Emory University and the director of the Living Links Center at Yerkes National Primate Research Center. De Waal studies the social behavior and cognition of monkeys, chimpanzees, and bonobos, focusing on cooperation, conflict resolution, and culture. His books include *Chimpanzee Politics, Peacemaking Among Primates, The Ape and the Sushi Master,* and, most recently, *Our Inner Ape.*

A native Texan, **David Dobbs** grew up in Houston, graduated from Oberlin College, and now lives in Montpelier, Vermont, with his wife and three children. He is the coauthor (with Richard Ober) of *The Northern Forest* (1995) and the author of *The Great Gulf* (2000) and *Reef Madness: Charles Darwin, Alexander Agassiz, and the Meaning of Coral* (2005). The son of physicians, he takes a particular interest in medicine, science, and how culture shapes our perceptions of natural phenomena — from forest ecology to synaptic networks — and our attempts to manipulate them. He contributes regularly to *Slate, Scientific American Mind, Audubon,* and the *New York Times Magazine,* and is currently working on projects about autopsy and fear. You can find more of his writing at http://daviddobbs.net.

Mark Dowie teaches science at the University of California's Graduate School of Journalism. He is the former editor-at-large of *InterNation,* a feature syndicate based in New York, and a former publisher and editor of *Mother Jones.* With the Massachusetts Institute of Technology in 2005, he began investigating the 150-year relationship between organized conservation and indigenous peoples; "Conservation Refugees" is a product of that study. During his thirty years in journalism Dowie has written over two hundred investigative reports for magazines, newspapers, and other publications and has won eighteen journalism awards, including four National

Magazine Awards and a George Polk Award, as well as citations from the National Press Club, Sigma Delta Chi, Project Censored, the University of Kansas (William Allen White Gold Medal), the University of Missouri (Penny-Missouri Award), and *The Best American Science Writing 2005.* In 1982 he was awarded the bronze medallion by Investigative Reporters and Editors (IRE), his fourth award from that organization. In 1992 Dowie received the Media Alliance's Meritorious Award for Lifetime Achievement, and in 1995 he was named a doctor of humane letters by John F. Kennedy University.

John Hockenberry, a three-time Peabody Award winner, a four-time Emmy Award winner, and a correspondent for *Dateline NBC,* has been a journalist and commentator for more than two decades. He has reported from all over the world in virtually every medium and has anchored programs for network and cable TV and for radio. He is the author of *Moving Violations,* a memoir, and *River Out of Eden,* a novel.

John Horgan is the director of the Center for Science Writings at the Stevens Institute of Technology in Hoboken, New Jersey, as well as a science journalist. A former senior writer at *Scientific American* (1986–1997), he has also written for the *New York Times, Time, Newsweek,* the *Washington Post,* the *Los Angeles Times,* the *New Republic, Slate, Discover,* the *Times* (London), the *Times Literary Supplement, New Scientist,* and other publications around the world. His three books include *Rational Mysticism* (2003), *The Undiscovered Mind* (1999), and *The End of Science* (1996).

Gordon Kane, a theoretical particle physicist and particle cosmologist, is Victor Weisskopf Collegiate Professor of Physics at the University of Michigan and the director of the Michigan Center for Theoretical Physics. He has published more than 170 papers on topics such as supersymmetry, Higgs bosons, string phenomenology, the matter-antimatter asymmetry of the universe, and the dark matter and dark energy of the universe. He has been a J. S. Guggenheim Fellow and is a fellow of the American Physical Society, a fellow of the American Association for the Advancement of Science, and a fellow of the Institute of Physics of England. He is the author of several books, two of which, *The Particle Garden* and *Supersymmetry,* are written for any curious reader.

Kevin Krajick is a New York City–based journalist specializing in science and the environment. A 1977 graduate of Columbia Graduate School of Journalism, he is the author of more than 250 magazine and newspaper articles for publications such as *The New Yorker, Newsweek, National Geographic, Science, Audubon, Smithsonian,* and the *New York Times.* He has voyaged on icebreakers, traversed glaciers in the Yukon and Peru, and climbed to the tops of the world's tallest trees (in the Pacific Northwest) to report on climate change and life in extreme environments. For stories on geology he has explored the Arctic tundra and crawled to the bottom of the world's deepest mines (in South Africa). He has been a finalist for the prestigious National Magazine Award for Public Service and is a two-time winner of the American Geophysical Union's Walter Sullivan Award for Excellence in Science Journalism. His book *Barren Lands: An Epic Search for Diamonds in the North American Arctic* (2001) tells of two small-time mineral prospectors who discover diamonds in Canada's Northwest Territories and of their impact on that remote region.

Robert Kunzig is a contributing editor of *Discover* and the author of *Mapping the Deep,* a book about oceanography. With the climate scientist Wally Broecker of Columbia University, he is writing a book about global warming, to be published in 2007.

Juan Maldacena, a professor in the School of Natural Sciences at the Institute for Advanced Study in Princeton, New Jersey, was in the physics department at Harvard University from 1997 to 2001. He is currently studying various aspects of the duality conjecture described in the article in this volume.

Charles C. Mann is the author, most recently, of *1491,* a history of the Americas before Columbus. A correspondent for the *Atlantic Monthly, Science,* and *Wired,* he has covered the intersection of science, technology, and commerce for many newspapers and magazines here and abroad.

Chris Mooney is the Washington correspondent for *Seed* and a senior correspondent for the *American Prospect.* He focuses on issues at the intersection of science and politics and is the author of the best-selling book *The Republican War on Science.* His next book, on hurricanes and global warming, is slated to appear in 2007.

Dennis Overbye is a science reporter for the *New York Times.* He is the author of *Einstein in Love: A Scientific Romance,* which was a Los Angeles Times Book Prize finalist, and *Lonely Hearts of the Cosmos: The Story of the Scientific Quest for the Secret of the Universe,* a finalist for the National Book Critics Circle Award and the Los Angeles Times Book Prize and the winner of the American Institute of Physics award for science writing. He lives in Manhattan with his wife, Nancy Wartik, and their daughter.

Paul Raffaele covered the world for *Reader's Digest* with award-winning reporting in the 1990s and is now a contributing writer for *Smithsonian.* His assignments have included an investigation of inherited slavery in North Africa; the African bush-meat crisis; a South Pacific cult that worships an American World War II soldier; a brutal rebellion in Uganda; and the last cannibal tribe on earth, treehouse dwellers in a remote New Guinea jungle.

Daniel Roth is a senior writer with the new Condé Nast business magazine, *Portfolio.* He has spent the last decade covering the biggest names in business and the innovators who aim to unseat them. Prior to joining Condé Nast, Roth was a senior editor at *Fortune,* where he guided the magazine's tech coverage after the dot-com bust. Roth lives in New York with his wife, Lisa, and son, Jack.

Jessica Snyder Sachs is the author of the upcoming book *Living with Microbes: Health and Survival in a Bacterial World,* to be published in 2007.

Oliver Sacks is a physician and the author of nine books, including two collections of case histories, *The Man Who Mistook His Wife for a Hat* and *An Anthropologist on Mars,* in which he describes patients struggling to adapt to various neurological conditions. His book *Awakenings* inspired the Oscar-nominated film of the same name and the play *A Kind of Alaska,* by Harold Pinter. He practices neurology in New York City.

David Samuels lives in Brooklyn and is a contributing editor at *Harper's Magazine.* His articles on fakes, frauds, demolition experts, network marketing gurus, hip-hop stars, greyhound racing fans, pineapples, and other natural and unnatural marvels also appear in *The New Yorker,* the *Atlantic Monthly,* and the *American Scholar.*

Josh Schollmeyer is an editor at the *Bulletin of the Atomic Scientists,* where he covers a wide range of global security issues. His film criticism has appeared in *Cinescape* and *Total Movie and Entertainment.*

Moncef Zouali completed his postgraduate studies in immunology at the University of Paris. As a postdoctorate research associate at Tufts University Medical School, he became interested in autoimmunity. He received a John Fogarty International Award from the National Institutes of Health and has been a senior Fulbright Scholar. In 1987 he returned to the Institut Pasteur in Paris to study the development of B lymphocytes in relation to autoimmunity. Currently he is a director of research at Inserm, the French National Medical Council in Paris. He has received national and international awards for his work on autoimmune diseases. In addition to publishing articles in professional journals, Zouali has edited several books, including *Molecular Autoimmunity.*

Other Notable Science and Nature Writing of 2005

SELECTED BY TIM FOLGER

BRUCE BARCOTT
As a Matter of Fact, Money Does Grow on Trees. *Outside,* March.
MARY BATTIATA
Silent Streams. *The Washington Post Magazine,* November 17.
WENDELL BERRY
Renewing Husbandry. *Orion,* September/October.

ADRIAN CHO
String Theory Meets Practice as Violinmakers Rethink Their Craft. *Science,* December 2.

PAUL DE PALMA
The Software Wars: Why You Can't Understand Your Computer. *The American Scholar,* Winter.
ROBIN DIXON
I Will Eat Your Dollars. *The Los Angeles Times,* October 20.
FREEMAN DYSON
The Tragic Tale of a Genius. *The New York Review of Books,* July 14.

MICHELLE FEYNMAN
The Feynman File. *Discover,* March.
MARIA FINN
Fulton Street Fish Market. *Gastronomica,* Summer.
RICHARD FORTEY
Blind to the End. *The New York Times,* December 26.

JOHN GALVIN
The Worst Jobs in Science. *Popular Science,* October.
LAURENCE GONZALES
The Biology of Attraction. *Men's Health,* September.
Hooked. *Men's Health,* October.

JEFF GORDINIER
The Fat Doctors. *Details,* January/February.
GORDON GRICE
Bite of the Hobo Spider. *Discover,* September.

JACK HANDEY
In Praise of the Human Body. *Outside,* May.

STEVEN JOHNSON
Dome Improvement. *Wired,* May.

VERLYN KLINKENBORG
What Do Animals Think? *Discover,* May.
JEFFREY KLUGER
Ambition: Why Some People Are Most Likely to Succeed. *Time,* November 14.
BRENDAN I. KOERNER
Blood Feud. *Wired,* September.
ELIZABETH KOLBERT
The Climate of Man. *The New Yorker,* April 25 and May 2.

DAVID LEONHARDT
Keeping Score: Baseball's Leading Man of Math Has Some Second Thoughts About the Numbers. *The New York Times,* April 24.
CHARLES H. LINEWEAVER AND TAMARA M. DAVIS
Misconceptions About the Big Bang. *Scientific American,* March.

JOSH MCHUGH
The Firefox Explosion. *Wired,* February.
BILL MCKIBBEN
Mad Max Meets American Gothic. *Orion,* November/December
JAMES MCMANUS
My Extremely Fancy Colonoscopy. *The Believer,* November.
LAWRENCE MILLMAN
The Delphic Fungus. *Orion,* January/February.
OLIVER MORTON
Life, Reinvented. *Wired,* January.

APURVA NARECHANIA
Hearing Is Believing. *The American Scholar,* Summer.
ERIK NESS
Detroit Is Still Stuck in Reverse. *On Earth,* Winter.
ANNALEE NEWITZ
The Conlangers' Art. *The Believer,* May.

H. ALLEN ORR
Devolution. *The New Yorker,* May 30.
LAWRENCE OSBORNE
Strangers in the Forest. *The New Yorker,* April 18.

DENNIS OVERBYE
On Gravity, Oreos, and a Theory of Everything. *The New York Times,* November 1.

TONY PERROTTET
The Semen of Hercules. *The Believer,* August.
THOMAS POWERS
An American Tragedy. *The New York Review of Books,* September 22.

LISA RANDALL
3-D Is Our Cosmic Destiny. *Seed,* October/November.
GARY RIVLIN
The Sniffer vs. the Cybercrooks. *The New York Times,* July 31.

MARY BETH SAFFO
Accidental Elegance. *The American Scholar,* Summer.
ROBERT SAPOLSKY
The Desert People Are Winning. *Discover,* August.
SETH SCHIESEL
World of Warcraft Keeps Growing, Even as Players Test Its Limits. *The New York Times,* February 10.
CHARLES SIEBERT
Planet of the Retired Apes. *The New York Times Magazine,* July 24.
MICHAEL STROH
Life Built to Order. *Popular Science,* February.
BRUCE STUTZ
Europe's Black Triangle Turns Green. *On Earth,* Spring
NEIL SWIDEY
What Makes People Gay? *The Boston Globe Magazine,* August 14.

NATHAN THORNBURGH
Unsafe Harbor. *Time,* October 10.

LINTON WEEKS
Count Him In. *The Washington Post,* May 23.
RICK WEISS
The Power to Divide. *National Geographic,* July.
MARGARET WERTHEIM
Origami as the Shape of Things to Come. *The New York Times,* February 15.
JEFF WHEELWRIGHT
Native America's Alleles. *Discover,* May.
FLORENCE WILLIAMS
Toxic Breast Milk? *The New York Times Magazine,* January 9.
DONALD WORSTER
A Long Cold View of History. *The American Scholar,* Spring.
KAREN WRIGHT
The Day Everything Died. *Discover,* April.

THE B·E·S·T AMERICAN SERIES®

THE BEST AMERICAN SPORTS WRITING™ 2006. Michael Lewis, guest editor, Glenn Stout, series editor. "An ongoing centerpiece for all sports collections" (*Booklist*), this series stands in high regard for its extraordinary sports writing and top-notch editors. This year's guest editor, Michael Lewis, the acclaimed author of the bestseller *Moneyball,* brings together pieces by Gary Smith, Pat Jordan, Paul Solotaroff, Linda Robertson, L. Jon Wertheim, and others.

ISBN-10: 0-618-47021-2 / ISBN-13: 978-0-618-47021-1 $28.00 CL
ISBN-10: 0-618-47022-0 / ISBN-13: 978-0-618-47022-8 $14.00 PA

THE BEST AMERICAN TRAVEL WRITING 2006. Tim Cahill, guest editor, Jason Wilson, series editor. Tim Cahill is the founding editor of *Outside* magazine and a frequent contributor to *National Geographic Adventure.* This year's collection captures the traveler's wandering spirit and ever-present quest for adventure. Giving new life to armchair journeys are Alain de Botton, Pico Iyer, David Sedaris, Gary Shteyngart, George Saunders, and others.

ISBN-10: 0-618-58212-6 / ISBN-13: 978-0-618-58212-9 $28.00 CL
ISBN-10: 0-618-58215-0 / ISBN-13: 978-0-618-58215-0 $14.00 PA

THE BEST AMERICAN SCIENCE AND NATURE WRITING 2006. Brian Greene, guest editor, Tim Folger, series editor. Brian Greene, the best-selling author of *The Elegant Universe* and the first physicist to edit this prestigious series, offers a fresh take on the year's best science and nature writing. Featuring such authors as John Horgan, Daniel C. Dennett, and Dennis Overbye, among others, this collection "surprises us into caring about subjects we had not thought to examine" (*Cleveland Plain Dealer*).

ISBN-10: 0-618-72221-1 / ISBN-13: 978-0-618-72221-1 $28.00 CL
ISBN-10: 0-618-72222-X / ISBN-13: 978-0-618-72222-8 $14.00 PA

THE BEST AMERICAN SPIRITUAL WRITING 2006. Edited by Philip Zaleski, introduction by Peter J. Gomes. Featuring an introduction by Peter J. Gomes, a best-selling author, respected minister, and the Plummer Professor of Christian Morals at Harvard University, this year's edition of this "excellent annual" (*America*) gathers pieces from diverse faiths and denominations and includes writing by Michael Chabon, Malcolm Gladwell, Mary Gordon, John Updike, and others.

ISBN-10: 0-618-58644-X / ISBN-13: 978-0-618-58644-8 $28.00 CL
ISBN-10: 0-618-58645-8 / ISBN-13: 978-0-618-58645-5 $14.00 PA

THE BEST AMERICAN GOLD GIFT BOX 2006. Boxed in rich gold metallic, this set includes *The Best American Short Stories 2006, The Best American Mystery Stories 2006,* and *The Best American Sports Writing 2006.*

ISBN-10: 0-618-80126-X / ISBN-13: 978-0-618-80126-8 $40.00 PA

THE BEST AMERICAN SILVER GIFT BOX 2006. Packaged in a lavish silver metallic box, this set features *The Best American Short Stories 2006, The Best American Travel Writing 2006,* and *The Best American Spiritual Writing 2006.*

ISBN-10: 0-618-80127-8 / ISBN-13: 978-0-618-80127-5 $40.00 PA

HOUGHTON MIFFLIN COMPANY www.houghtonmifflinbooks.com

LINDA
JIM

I AM WATCHING

Also by Emma Kavanagh

The Missing Hours